The Psychology of Flight Training

The Psychology

of Flight Training

Ross Telfer

John Biggs

Iowa State University Press / Ames

Ross Telfer is Associate Professor and Director, Institute of Aviation, University of Newcastle, New South Wales, Australia.

John Biggs is Professor of Education, University of Hong Kong.

Composed by Iowa State University Press
Printed in the United States of America

First edition, 1988

Library of Congress Cataloging-in-Publication Data

Telfer, Ross, 1937–
The psychology of flight training/Ross Telfer, John Biggs.—1st ed.
p. cm.
Includes index.
ISBN 0-8138-1347-6
1. Flight training—Psychological aspects. I. Biggs, John B. (John Burville) II. Title.
TL712.T43 1988 87-29952
629.132'52'071—dc 19 CIP

Contents

Foreword **vi**

Preface **vii**

Acknowledgments **ix**

1 What's Involved in Learning to Fly an Aircraft? **3**

2 Learning and Memory **14**

3 Teaching the Skills of Flying **44**

4 Motivation and Arousal: General Determinants of Behavior **58**

5 Motivation and Self-concept of Student Pilots **78**

6 Evaluating Learning **94**

7 Aspects of Training and Instruction **119**

8 Conclusion **147**

Glossary **151**

Index **161**

Foreword

In this book, Ross Telfer and John Biggs bring off a small miracle. They smash the mold of the "recipe" flight instructor through a process of psychological analysis followed by logical synthesis. Like magic, we have a new, easily understood mold of the business of teaching others to fly, a sleight-of-hand transition from the science of learning to the professional art of flight instruction.

STAN ROSCOE

ILLIANA Aviation Sciences
Las Cruces
New Mexico

Preface

IN THE LATE SEVENTIES, the authors were given the opportunity to design a short course aimed at providing flight instructors with a professional background in teaching knowledge and skills. Lacking precedents to work from, their starting point was experience: a few flying lessons later, the program began to take shape.

The program included aspects of the psychology of learning, videotaped teaching models that identified essential skills and their application to flight-related topics, the opportunity to develop and utilize aids to instruction, and practice in the use of audiovisual aids. The emphasis was on the instructor as a catalyst in the process of learning. The production of this book has been a parallel exercise.

Since that first instructor course was introduced, a similar form of program has become mandatory for flight instructor licensing. A simple approach to the preparation of this book would have been to produce a primer for the flight instructor licensing requirements. That would have been the approach most likely to meet with market success. However, there are three good reasons why that approach was not adopted.

First, flight instructors are professionals. They are not trained simply how to act, but are educated so that they are aware of the reasons for acting in certain ways and not in others. Flight instructors have the same occupational characteristics as other professionals: their craft has a technical basis, they have an exclusive jurisdiction to practice that craft, and there are prescribed standards they have to maintain. A "recipe" approach is completely inconsistent with such professionalism.

Thus in this volume the reader will not find simple solutions to the complexities of flight instruction and evaluation. Professional autonomy involves the ability to make such a choice from a number of alternatives. On the day, with the learner, for the particular exercise, in the prevailing conditions, for the individual instructor, there will be an optimal alternative. Only the professional can choose.

A second reason for avoiding the recipe approach is that it implies there is an easily identifiable body of knowledge that is the only appropriate preparation an instructor requires. In reality, of course, the situation is quite different. Each student is different, each instructor is different. The time of the day, the state of the weather, the experience of the previous flight—all are variables the instructor considers before the first words of the day. The flight instructor needs as rich a preparation as time will permit, and time covers a full career. The most successful teachers in a variety of fields agree that one never stops learning or refining one's skills.

Experienced flying instructors are uniquely skilled individuals. Many of them have distilled years of contact with students to provide their personal reservoir of alternatives to meet individual needs. For them, this book may provide the realization that what they have been practicing intuitively has a firm foundation in psychological theory. For the beginning instructor, however, this book offers a means of acquiring some of the principles and rationale that previously only long hours of experience had brought.

Our third reason for avoiding the recipe approach is because it is very similar to the traditional molding of flight instructors. In this process, successive generations of instructors have taught people to fly in the same ways as they themselves learned. There appears to have been a tendency to blur the distinction between a highly skilled pilot and a highly skilled teacher. Because of the unique combination of experience, skill, knowledge, and values that make a top pilot, there may still be a tendency to defer to flying ability rather than teaching ability. This is not an argument where one can have the latter without the former: it's a call for parity of esteem. Both are needed for a dynamic aviation industry. Teaching people to fly requires top-level skills in both flying and teaching.

We have seen our role as a link between the reader and those aspects of the process of learning that we regard as important in our understanding of instruction, check and training in aviation. Thus, there are few footnotes or references to impede the flow of the text. For those interested in follow-up of particular aspects, a brief reading guide to key sources follows each chapter. This book is designed for the professional, not the academic.

A final point can be made about the structure of the book. We see it as a bridge between learning theory and instructional practice. We have been in the position to provide the first, and we have been fortunate to receive assistance to gain our examples of the second. Requests for assistance from instructors, chief pilots, check and training captains, and student pilots always resulted in valuable insights. To that extent, this volume is also a production of the aviation industry.

The focus of this book is on the application of the process of learning to the methods used in flight instruction, check and training. Where the actual content of instruction is incorporated, it is used only to illustrate the method: not to provide the content for a reader already expert in the area. The emphasis in the chapters that follow is upon the process of instruction, not upon the content of that instruction. For that reason, many of our international examples have been left in their original form, regardless of their applicability in other countries of the world where this book may be read.

ROSS TELFER / JOHN BIGGS

Acknowledgments

AS A PUBLICATION in a very narrow field of specialization, this book came to print only because of the interest and support of a number of key individuals in the aviation industry. Essential contributors were Jim and Maureen Spark of the Civil Air Training Academy at Cessnock Airport, who published the Australian edition; Stan Roscoe, ILLIANA Aviation Sciences, who provided helpful advice and considerable assistance; Dr. Al Diehl, U.S. Air Force Safety Center, and Prof. Dick Jensen, Ohio State University, who were encouraging and extremely helpful with pilot judgment training publications; and Sharon Field, who did the art work.

A number of professional pilots and aviation experts contributed examples of flight instruction, especially in RPT operations. Ray McNamara, Australian Department of Aviation, provided the general aviation instructional scenario forming Chapter One; three Australian airlines, QANTAS (Director of Flight Operations, Capt. A. I. Terrell; Operations Training Manager, David Cormack; and Dennis Christley, Line Operations Manager), Ansett (Capt. Ken G. Patton, Check and Training Captain), and East-West (Capt. Roger Miller, Chief Pilot), assisted in the provision of applications to commercial pilot practice.

A number of organizations, such as the National Association of Flight Instructors and Prentice-Hall, Publishers, permitted us to adapt text or illustrations. In order to simplify the content and to maintain flow, we have distilled the efforts of many of our academic colleagues and fellow authors. We gratefully acknowledge their contributions.

The Psychology of Flight Training

1 What's Involved in Learning to Fly an Aircraft?

CHAPTER ONE answers the following questions: What are three of the major elements of flight training? What are the major teaching skills in flight instruction? How can educational psychology be related to flight instruction? Is there a place for theory in the practical matter of teaching people to fly aircraft?

After reading this chapter, you should be able to (1) identify content, method, and rationale of a unit of instruction; (2) describe how questioning is a means of evaluating a student's progress; (3) show how the preflight brief is related to the flight; (4) identify at least three major sources of student pilot learning of attitudes and skills; and (5) apply the "What? How? and Why?" analysis of a flying lesson.

The Process of Flight Instruction: A Case Study

This chapter looks at Martin, a typical student, who has arrived for his weekly flight at the local flying school. Martin has had about seven hours of dual flight experience. This included lessons on the effects of controls, straight and level flight, turning, climbing, and descending. In his last flight, Martin was assessed by the instructor, who felt that it was time for an introduction to stalls. On the return to base that day the instructor had demonstrated a gentle stall. This Sunday's lesson is on recognition of the stall and standard stall recovery.

Martin was told to use the intervening week to prepare for the next lesson. Specifically, he was asked to investigate the causes of aerofoil stalling, the relationship between the angle of attack and the center of pressure, and the factors that affect the stalling speed of the aircraft. As he had done in the past, the instructor also made a list of some aspects that Martin had to learn by heart: the symptoms of the approach to the stall, the symptoms of the stall itself, the steps in the standard stall recovery, the stalling speed of the aircraft in various flap configurations, the stalling speed in clean configuration and straight and level flight, and the prestall safety checks.

The instructor is quite familiar with the chief flying instructor's notes on this air sequence and intends to follow them in this lesson. These notes are provided in Insert 1.1.

Here comes Martin now, and the instructor is waiting for him.

INSTRUCTOR: Good morning, Martin. How are you today?
MARTIN: Fine, thanks. Weather looks good, too.
I: Right. Now, Martin, the purpose of the exercise this morning is to

INSERT 1.1. Introduction to Stalling

Aims

This lesson aims to teach three aspects of stalling:

1. Recognition of the symptoms of the approach to the stall
2. Recognition of the symptoms of a stall
3. Recovery from a stall by standard method

Additionally, these procedures will be carried out so that the student is not left with apprehension about either stalling an aircraft or recovering it.

While all stalls in this exercise will be conducted from straight and level flight, the preflight briefing will include reference to stalls in other configurations and speeds.

Objectives

By the end of the instructional period, the student will be able to:

1. Accurately identify in sequence the symptoms of the approach to the stall and the symptoms of the stall
2. Demonstrate a standard stall recovery with an altitude loss of less than 200 feet

Preflight Briefing Procedure

1. Motivation/reassurance of student (aircraft easily recoverable by correct technique)
2. Objectives
3. Check essential knowledge:

   –stalling speeds from straight and level in clean and flapped configurations
   –prestall checks
   –symptoms of approach to stall
   –standard stall recovery

4. Overview of flight

examine stalling. Last time we flew I showed you a stall . . . nothing magic about it . . . quite gentle, really. And the important thing is that it was quite easy to recover from. Agreed?

M: Well, it certainly appeared that way.

I: If you go about it the right way, Martin . . . I guess that's the important point. You didn't seem to be all that uncomfortable or unhappy.

M: No, I wasn't uncomfortable.

I: Good—so we'll go into it a little more this morning. Let me remind you why we practice stalling and teach you to carry out the standard stall recovery. At various stages of flight we have to fly close to the stall condition. When, for example?

M: The approach . . .

I: The best example, Martin, sure, the approach . . . because we are approaching at a speed and in a configuration that gets close to the stall. We never want to unintentionally stall the aircraft. If we do, we want to be able to recover quickly without loss of altitude—which we don't have a lot of when on final approach. Our objective today is to recognize when the aircraft is close to the stall and to be able to recover an aircraft that has stalled. So we need to be able to recognize not only the symptoms of the approach to the stall, but also the symptoms of the stall itself. By the end of today's lesson you'll be able to carry out the standard stall recovery.

Let's pause to examine what's happening in this lesson, and what's happening in flight instruction generally. The instructor has established the *aim and objectives for the lesson.* Because of a training syllabus and his professional experience, the instructor is aware of what has to be taught and the sequence in which it is placed.

Notice the *technique of instruction?* The instructor is using questions to involve the student and to explore the student's background, knowledge, ability, and attitudes as they go into the lesson. This will enable ongoing and concluding evaluation to be made of any changes that have occurred as a result of the lesson. Simultaneously, the instructor is encouraging or motivating the student. Causes of anxiety are being removed or minimized. Rapport between the two is being strengthened and there is evidence of empathy in which the instructor can sense reactions or feelings. The student is being led from the known into the unknown.

Let's rejoin the lesson, still in its preflight briefing stage.

I: Martin, you should remember the checks I carried out before I conducted the stall when we last flew.

M: Yes . . . you asked me to learn them for today. Want to hear them?

I: How do you remember them . . . the way I told you?

M: Yes . . . the mnemonic A SEAL.

I: Good. That makes it easier. Let me hear them.

M: A is for altitude . . . over 3000 feet . . . S is for security . . . no loose objects in the cockpit.

I: Why?

M: They could be flying around and cause damage or accident.

I: They certainly could . . . good . . . anything else in security?

M: Yes, check seat and harness, doors locked . . . that's security.

I: Back to altitude . . . 3000 feet?

M: Minimum recovery height.

I: Yes, but above what?

M: Ground level.

I: Yes, good, that's the important point. We could be at a spot where there's an altitude of 1000 feet at ground level. Recovery is by 3000 feet above ground level.

M: OK. Now where was I? Oh, A, altitude, S, security. E, engine . . . Electric fuel pump on, mixture rich, check temperatures.

I: Yes, we'll go from power to no power . . . and don't forget temperatures and pressures . . . sufficient fuel for the exercise . . . and then?

M: A is for area . . . because we have to be covering the training area where we'd have an alternate landing spot selected. And L is for lookout. We'll do a 360 to ensure the area is clear . . . checking above and below, too.

I: And we'll do the lookout before every stall. Right, A SEAL checks complete. Now think back to the stall last week. How did I describe the configuration of the aircraft as I brought about the stall?

M: We were straight and level, no flaps, no power. Clean, wasn't it? Yes, you said we were in the clean configuration.

I: That's right, Martin. After carrying out the checks, all I did from straight and level was to close the throttle and maintain altitude. How did I do that?

M: Raised the nose?

I: Good, and . . . ? Power came back, remember?

M: Left rudder.

I: Right on. Then came the symptoms of the approach to the stall. Some more homework, Martin, what are the symptoms of the approach to the stall?

M: Airspeed reduced, higher nose attitude, sloppy controls.

I: And?

M: Shuddering, wasn't it?

I: Even before . . . the mechanical aid we heard . . .

M: Oh—the stall warning indicator.

I: Yes. It comes in before the shudder, which we call "buffet." So we have quite a few symptoms before we enter the stall: decreasing airspeed, higher nose attitude, sloppy controls, stall warning, buffet. Give them to me again, Martin.

M: Decreasing airspeed, high nose . . . ah . . . sloppy controls, stall warning, buffet.

I: Good, but don't confuse those signs with the stall itself. What does the buffet indicate, Martin? Aerodynamically?

M: Loss of flow over the wings . . . loss of lift.

I: Well, we're getting breakaway . . . turbulent air flow from the wings flowing over the elevators. The next step is the stall . . . what are its symptoms?

M: Nose pitches down, loss of altitude, possible wing drop.

I: The nose can pitch down; it may pitch back up . . . but you can see the loss of altitude on the performance instrument: that tells you that the aircraft is stalled. As a built-in safety feature, our aircraft is quite mild in the stall; but some you fly later in your career may be different. Right—we're in the stall. Now what?

M: Get flow back over the wings . . . get some lift.

I: Not chapter and verse—what are the action steps you learned?

M: Simultaneous control column forward and power on.

I: Good. What about the wings?

M: When the aircraft is unstalled, level wings with rudder . . .

I: Good. Look at the horizon to level the wings with rudder. Now what?

M: Regain altitude.

I: Yes. Good. I'll show you the attitude that will enable us to climb away as effectively as possible to regain the altitude we've lost. OK?

M: OK.

I: Why do you think I've been making such a big deal out of your knowing all these checks, symptoms, and action steps before we even get in the aircraft?

M: I guess it leaves flying time for flying.

I: Right! And you've shown that you have the essential knowledge. So let's plan our flight. I want you to take us out to the training area, climbing to 4000 feet. Set the aircraft up in straight and level. I'm not going to say a word between now and then.

M: No problem.

I: Of course not, you did it last week. In the training area we'll do some turns, climbs, and descents before we go into the stalling exercise. I'll demonstrate, then talk you through, and by the end of the exercise I'll be able to sit back and say, "OK, Martin, show me a stall in clean configuration from straight and level. Recover using the standard technique."

M: Sounds fine.

I: Then, Martin, we'll come back and shoot a couple of traffic patterns. Any questions?

M: No, all clear.

I: Let's go.

M: Will I do the preflight check?

I: As I said, you do everything. You won't hear from me until we're at 4000 feet.

The preflight brief is now complete. What has happened? Both instructor and student clearly understand the purpose of the lesson, and they both know the criterion by which it will be evaluated. At the end of the lesson the student will be asked to show his ability in a stalling exercise under specified conditions (straight and level, clean configuration) with a specified level of performance (altitude loss).

The procedure has been spelled out. Both instructor and student share the same expectations of the lesson. The steps are clear and there are common goals.

Note, too, the ways in which the student gains knowledge, attitudes, and skills. A variety of sources is being utilized. (See Fig. 1.1.) The student has read in the area, has been given further information from the instructor, has discussed and clarified his views, and can use the instructor as a model.

There are also some principles of instruction at work in this lesson. The instructor is working from what the student knows (or has experienced) to the unknown, from the concrete to the abstract, and from specific examples to general applications.

Let's return to the cockpit and monitor the flight sequence for a few minutes.

I: Good, Martin. You're holding 4000 feet quite nicely. Is it trimmed?

M: Yes.

I: Just take your hands and feet off . . . good. You're getting near the edge of the training area. Do a medium turn to the left, rolling out when you're pointing at Mount Sugarloaf. Good. Now do the same to the right. OK. Well done. Maintain the heading and enter a glide descent.

M: Power coming off.

I: Whoa, there! Taking over.

M: Handing over.

I: What's the first thing we do before entering the descent?

M: Check?

I: Yes—we must clear the area below us. Watch me and follow through. Right? Handing over.

M: OK. Taking over.

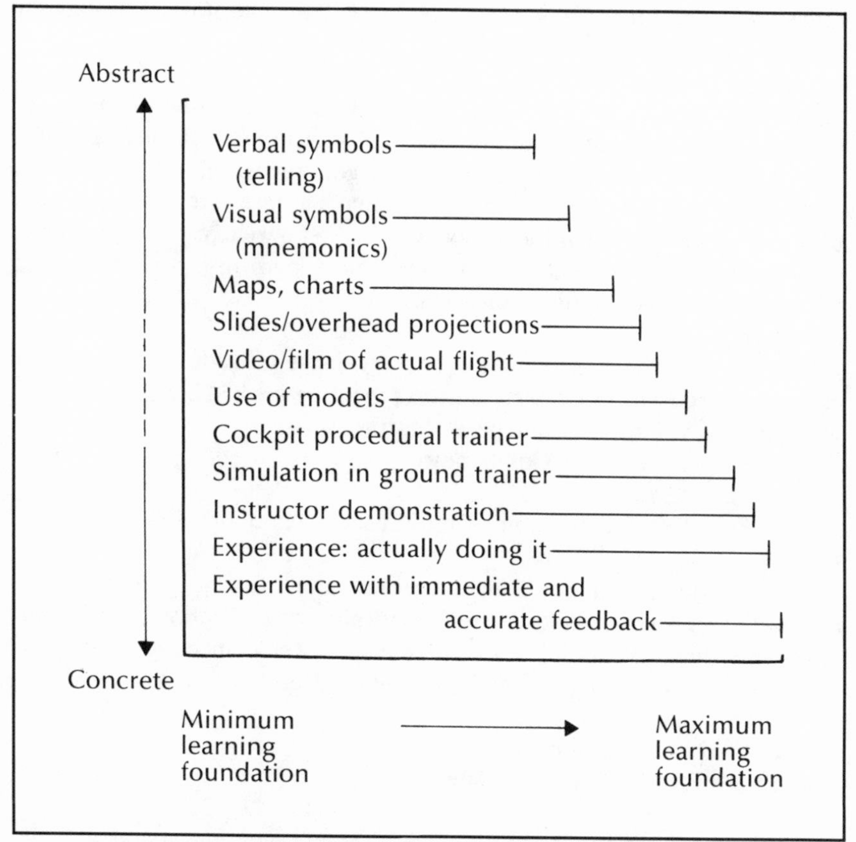

FIG. 1.1. **Sources of Knowledge, Attitudes, Skills**

I: Now check, and if clear, descend to 3000 feet maintaining this heading. Good. Now a medium, level turn to the right to get us back into the training area, rolling out when we are heading for Mount Sugarloaf. Fine—now let's get back up to 4000 feet, Martin. Maintain the heading, enter a standard climb leveling off at 4000 feet. That's just what we need. Good. Let's try some stalls. First step?

M: A SEAL checks.

I: Right. I'll fly the aircraft while you go through them.

M: Handing over.

I: Taking over.

(Student completes checks.)

I: Now sit back, watch, and listen. Look outside unless I direct your attention inside. I'm bringing the power back gently, preventing yaw with a touch of left rudder. As power comes back I lift the nose. Look at the airspeed. What's it read?

M: 70 knots.

I: OK—first symptom, decreasing airspeed. Notice how I continue to raise the nose to maintain altitude. Second symptom, high nose attitude. Feel the controls . . . sloppy . . . and there goes the warning horn . . . and now comes the buffet. Here we go, Martin . . . nose pitching down, up, back down. Column right back. Look at the VSI—we're certainly descending. Check the altimeter. This is the stall. No wingdrop this time . . . and not all that uncomfortable, either, is it?

M: No.

I: Now for the standard stall recovery. Nose attitude. Power coming on—wings are level so I can start easing the nose back. I'll hold that while we climb away. How's the airspeed, Martin?

M: Increasing—on 80 knots now.

I: Altimeter?

M: 3600 feet and climbing. We must have lost about 400 feet. Is that usual?

I: No. Remember how I held it in the stall? Just goes to show how much altitude you could lose, though. Ideally, we'll quickly recognize the symptoms of the stall and react immediately. Remember the symptoms? Speed coming off, nose attitude high, sloppy controls, stall warning horn, buffet. Any questions?

M: Will I hold it in the stall, too?

I: Let's take it a step at a time.

The remainder of the exercise follows the sequence as planned.

Elements of Flight Instruction

The flying sequence draws attention to the art as well as the skill of instruction. The way the instructor draws attention to aspects of airmanship, the grasped opportunities to spontaneously provide the knowledge, attitudes, or skills of a competent pilot, and the systematic development of the planned lesson—all are characteristics of the experienced flight instructor.

In essence, this briefing and exercise on stalling shows the elements of flight instruction:

1. *Content:* prescribed to some extent by a training syllabus, but ultimately determined by the professional judgment of the flight instructor (as guided by the chief flying instructor). The selection and sequencing of

content can be seen as a crucial component.

Flight instructors must decide, for each of their students, the next step in learning to fly. Additionally, this occurs both in terms of a series of lesson units and in terms of planning the content of the lessons themselves. Thus any analysis or flight instruction must refer to both programming and selecting lesson content.

2. *Technique:* consists essentially of a range of teaching skills. Educators are able to identify the major skills involved, and to provide advice on ways flight instructors can implement successful teaching methods. Such skills include reinforcing or motivating students, varying lessons to maintain student interest, questioning techniques, beginning and ending lessons, explaining, using teaching aids, and so on.

3. *Rationale or justification:* these methods and approaches can be found in educational psychology. How people learn and remember provides assistance for flight instructors who wish to help their students learn to fly as effectively and as efficiently as possible.

Asked for the rationale for their approaches to flight instruction methods and programming, most instructors would probably respond "Because it works." If you think about it, information that enables us to predict what will work and why it works is termed "theory." Although many instructors would probably deny it most strenuously, methods of flight instruction are based upon theory.

On the basis of experience or observations, instructors predict that their students will react in a certain way. By following a sequence of flying exercises, instructors know that they can solo a student in 15 hours. By following another approach, they know how to put an anxious student at ease or develop self-confidence in another.

Logically, one cannot argue that there is no place for theory in flight instruction. At best, one can argue that there is no place for theory without application. This book aims to provide principles and advice upon which instructors can act. It does not presume to offer a recipe because the authors recognize individual differences: an endless diversity of both instructors and students. They all differ, and the ideal mix between instructor and student is a highly complex matter.

The recipe approach is also inappropriate because the menu differs. Managers of airline flying standards report a constant check and training problem in the initial training phase, due to the variation of pilots' previous experience levels, for example, adapting to the disciplined, coordinated, crew operating concept. The training structure caters to these variations through individualized evaluation and appropriate remediation.

In a changeover from F27s to F28s, further training needs were isolated. Tasks had to be performed within a shorter time span. The tempo of the operation is continuous, particularly in the short-haul environment. Instrument scan rates had to be increased. Achieving operational efficiency required even greater application. Handling techniques also had to be modified to suit the new characteristics of the aircraft.

Airlines have some inherent training problems. For example, an RPT pilot normally encounters few landings per flying hour. To postulate a prescriptive approach to flight instruction, as well as check and training operations, is patently absurd.

It is possible, however, to provide guides to enable instructors to make optimal choices when deciding lesson content and teaching technique. These guides come from a consideration of the three essential elements that can be found in the lesson on stalling:

1. WHAT was taught (content)
2. HOW it was taught (method)
3. WHY it was being taught in that way (rationale)

These elements and their associate guides for the instructor can be extracted from available knowledge in the fields of the psychology of learning, instructional skills, instructional aids, and instructional design. The following chapters deal with these aspects as they apply to flight instruction.

Summary and Conclusion

An instructor is rarely in the position of having to make all of the necessary decisions about the content of lessons for a student. A prescribed syllabus, from higher authority or regulation, usually dictates what is to occur. The instuctor has important responsibilities, however, in terms of the teaching technique that is employed. This technique is based upon rationale, a theory, justifying its choice.

In the next chapter we look at the stages of the learning process, and the ways people learn to fly an aircraft.

Discussion Questions

1. From a training syllabus, select and name an exercise. Briefly describe the method you would use for the exercise, and then present its rationale. Do not provide the content.
2. Now refer to Fig. 1.1 and analyze the extent to which your approach was predominantly abstract or concrete. To what extent does this analysis fit your rationale? What is your evidence? What is the link with the rationale?

3. For the exercise discussed in 1 and 2 above, write one aim and one objective to clearly indicate the difference between the two and give the rationale.
4. What are three major sources of student pilot learning of attitudes and skills?

Further Reading

The Center for Vocational Education. *Professional Teacher Education Module Series.* American Association for Vocational Instructional Materials, University of Georgia, 1978. A series of 100 booklets focusing upon the professional competencies of vocational teachers. Each booklet integrates theory and application and provides a criterion-referenced assessment of the reader's performance of the specified competency. Relevant to this chapter are the six modules on instructional planning: determining student needs and interests, developing student performance objectives, developing a unit of instruction, developing a lesson plan, selecting student instructional materials, and preparing teacher-made instructional materials. AAVIM will provide details of availability and prices of all or any of the booklets in the set.

Curzon, L. B. *An Outline of Principles and Practice: Teaching in Further Education.* 2d ed. London: Cassell, 1980. Part 2 is on communication, control, and teaching objectives, including chapters on the definition and construction of teaching objectives, and taxonomies of teaching objectives.

2 Learning and Memory

THIS CHAPTER details how we process and store information. There are factors, controllable by the flight instructor, which determine whether or not a student will pay attention. There are specific instructional strategies that make it more likely that the learner will learn and that the information presented in the preflight brief will be remembered.

The following questions are answered in this chapter: What are the three broad stages of the learning process? How many memory systems are there? Why do we pay attention? How does a communication gap occur? What instructional strategies direct student attention? How much information can we process? Do students learn by their mistakes? What is the difference between long-term and short-term memory? What factors are involved in forgetting?

After reading this chapter, you should be able to (1) help a student identify important aspects of a flight sequence, (2) choose from five instructional strategies for influencing a student's mental set to learn, (3) present information to a student in a way that will facilitate remembering, and (4) instruct in a way that will minimize forgetting important aspects of the lesson.

Three Stages of Learning to Fly

When humans learn any complex task, it is helpful to distinguish three broad stages: (1) attending to the particular stimuli in an environment busy with activity, (2) processing the information presented by the selected stimuli, and (3) storing it so that it may be used later.

ATTENDING

In the flying situation we experience a huge variety of sensations. The number of things we *could* pay attention to is impossibly large. Let us consider a moment. Visually, we may attend to external reference points on the horizon or traffic in the traffic pattern, or there may be key indicators on the instrument panel. A stall warning may sound, and g forces could be at work on our bodies. Now from all these sources transmitted through our five senses, a selection is, in fact, held and registered very briefly. We have to select from this sensory register to determine what we attend to, as we are distinctly limited in the number of things we can do at once. For example, it is not possible to read an aircraft manual and listen to a weather forecast at the same time, nor attend to an instructor's conversation while daydreaming. We can attend selectively to only one train of thought at a time.

Consider the dilemma of the relatively new first officer who has a busy time coming up with an impending approach for landing. The door of the flight deck opens, bringing a slight but unmistakable fragrance of after-shave to the cockpit. It's an attractive male cabin attendant with a request for a passenger car-hire reservation radio-call. Now there's real competition for sensory input! To which sensation will the female first officer attend?

To redress the potential problems of the situation, it is necessary to recognize and maintain priorities. This becomes even more necessary when a first officer is seen as a captain in training in the restricted teaching space of a cockpit. The instructor's enjoyment of garlic in a meal the previous evening has to be balanced against the effect on the learner who may give greater priority to avoiding an offensive odor than listening to instructions. Similarly, the smell of fuel may preoccupy a student on first solo to the point that other vital inputs do not register.

PROCESSING

When we make up our minds to attend to something, we have to do something with that information. We think about it. To be more specific, we can *rehearse* it, by repeating it over and over again, or *code* it, by linking it to something we already know.

After a first flight, someone could remember the aircraft by repeating "Piper Cherokee, Piper Cherokee, Piper Cherokee" again and again, that is, by rehearsal. Second, one could use coding, by thinking "Cherokee, ah yes, the Indian tribe." This second way of remembering is usually better, but problems can arise: was that plane a Piper Comanche—or even a Piper Arrow?

Whether we code or rehearse, however, it is done consciously in working memory.

STORING

After processing, we need to store the information out of consciousness in long-term memory in such a way that it can be recalled to consciousness when required in the future.

Each of these stages requires that the information be held long enough for the process in question to be carried out; and each refers to three levels of memory.

Three Memory Systems

The three stages of attending, processing, and storing involve very different time scales. The act of selecting what to attend to requires that information be stored for up to one second in the sensory register. Working memory holds material for about a minute, unless the material is processed by coding or rehearsing it, when the material may be retained in long-term memory for hours, years, or even a lifetime.

These time scales make a convenient framework for the present discussion: ultrashort, short, and long-term. The three stages are shown in Fig. 2.1.

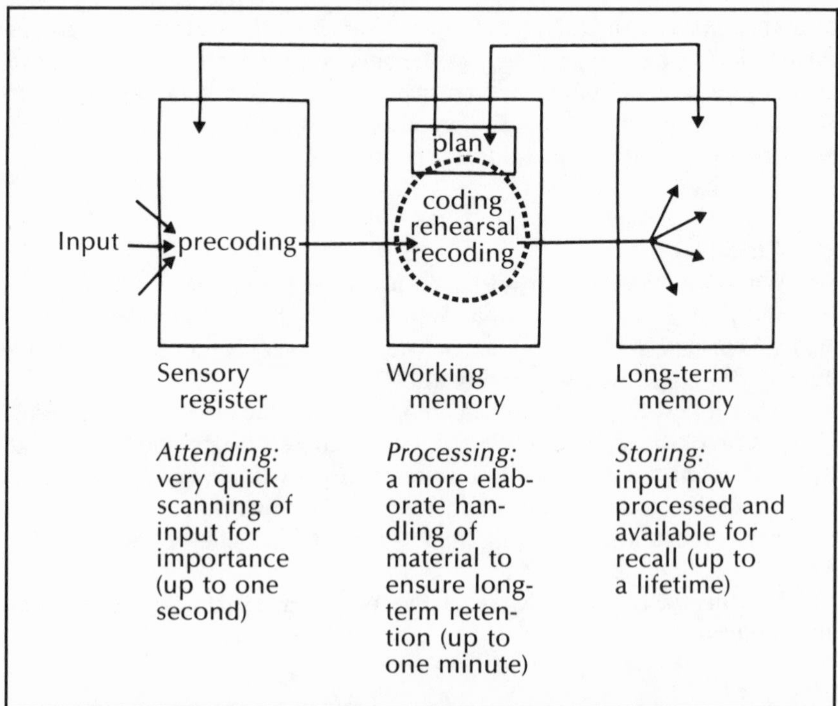

FIG. 2.1. **Three Stages in Learning and Memorizing**

We may think of each stage as a three-part system. The sensory register is where selective attention is carried out by a process called precoding. Working memory contains a plan, which governs both precoding and the processes carried out in working memory itself, such as coding, rehearsing, and recoding. The plan is assembled momentarily from long-term memory, and is in a continual state of reshuffling. After processing in working memory, material is then stored, out of current awareness, in long-term memory.

Let us now examine each of these stages in more detail.

THE SENSORY REGISTER

Ultrashort-term storage is carried out in the sensory register. All sensory impressions are retained for a period lasting up to one second, depending on several factors, such as the strength of the stimulus. We know from research that people can retain large numbers of digits for about one-third of a second, after which the retention rate drops dramatically to about seven or eight digits when the digits move from the sensory register to working memory.

The properties of the sensory register may lead instructors to believe they have the attention of their students when in fact nothing could be further from the truth, as the following example shows.

John is gazing out of the window, apparently miles away. The instructor knows that he is not attending. With a quick question he knows John will be caught:

Instructor: So there are three basic aspects in the safe recognition of and recovery from stalls: angle of attack, stall recognition clues, and recovery techniques. What are those three basics, John?

John: Angle of attack, stall recognition clues, and recovery techniques.

The instructor was sure the student was daydreaming, yet John answered correctly. John, in fact *wasn't* paying attention to the preflight brief as suspected. The problem was that he had been given a sensory register test, not a probe of his attention-paying, conscious, working memory. John, on hearing his own name, simply hauled in the jumble of words that was still echoing around in his sensory register and read them off. He could play it back, but that doesn't mean he was paying attention and understood that jumble of words.

The cocktail party, club lounge, or hotel bar can illustrate more properties of the sensory register. We can be in a crowded situation conversing with only a small group and conscious of the discussion within that group alone. Yet if a multidirectional microphone was placed where we stood, it would pick up a jumble of confused noises. My personal microphones are obviously selective, attending to one "frequency" and tuning out the others. If I hear my name mentioned in the group behind me, I can quickly change frequency. Without altering my stance or expression, I tune into that conversation, leaving the first one. I can change "frequencies" by conscious control.

When I was participating in one conversation, how could I be aware that my name was being mentioned in another? Probably because all the information that I could hear was being received at a level below my actual awareness. It was retained just long enough for a quick scan to decide which parts were worth attending to. Consciously, I was aware only of the result of that process when I became alerted to my name.

Why We Pay Attention

Three factors determine whether or not we pay attention to something: mental set, physical properties of the stimulus, and physiological states.

1. Mental set. Mental set is often established by a deliberate plan, for example, to read this page, to listen to the radio, or to talk to a friend. It is a conscious intention to attend to one line of activity rather than other possible ones.

If this is all there is to it, life would be easy. Concentration, however, can slip from one conversational train to another; or with the best will in the world, we simply go off daydreaming when we really should be listening to what someone is telling us. John undoubtedly recognized the importance of a knowledge of stall characteristics and recovery methods, but he still lapsed into a lack of concentration on the topic.

2. The physical properties of the stimulus. Physical properties of the stimulus make a difference. "Low key" stimuli are hard to attend to: a flat monotonous voice, a dimly lit picture, a blurred or opaque image. Bright lights, loud sounds, or variations of these are easy to attend to. *Variation in the stimulus and changing stimuli hold our attention very effectively.* Effective instructors deliberately vary the pitch, volume, and pace of the delivery of their subject.

When it becomes difficult to make such variations, one quickly becomes aware of the consequences. For example, at an instructor refresher course one of the authors was faced with the task of presenting the first session after quite a good lunch. The session was conducted in a room with an excellent view from large picture windows that overlooked a green landscape. It was summer, and the room lacked air-conditioning. When electric fans were switched on, they proved to be quite noisy. The presentation was based on a series of overhead projections, but the screen was located adjacent to the only door, which proved to be the one source of ventilation.

If the door was closed, the temperature became distracting. If the door was open, the effect of the light reduced the legibility of the screen. The fans were too noisy for audible speech. There was no happy ending, as the competing stimuli were too powerful.

3. Physiological or internal states. The third factor determining whether or not we pay attention is that of our personal state at the time. In an incipient spin, the student pilot may be more concerned about a feeling of anxiety, stress, and discomfort than the observation of the instructor's use of power and controls.

Outside stimuli that are threatening detract from performance. Other distractions range from the previous example of the flight attendant entering the flight deck to the aroma of brewing coffee or the smell of cooking when a coffee break or meal is overdue.

These three conditions add up to **IMPORTANCE.** Each one spells out a way in which external stimuli may be very important to us. Things can be important because we decide it, because they are physically insistent, or because they have biological survival value. A student pilot may have an urgent need to empty bladder or bowels; a first officer, having failed a first line check for command, is facing the second and final opportunity; or the unexpected rush of noise on takeoff as a door accidentally opens—all affect the way the individual will sort out what is seen as the important stimulus. This sorting of important from unimportant stimuli is called *precoding.*

A Communication Gap

For the student pilot, going solo or passing a license test may be priorities. What that student rates as important may be quite different from what the instructor sees as important at that stage. Another student, even another instructor, may have different priorities again.

Our different backgrounds, personal needs, and expectations make it probable that we will have different priorities and, quite literally, see a different world. If we work on different data, we are bound to disagree.

Instructing is a communication situation like any other, and so for communication between instructor and student to take place, it is necessary for each to have similar plans to construct a mental set so that each may precode the same information as important. In that way we attend to the same things.

On their solo flights, students could employ plans that are quite incompatible with those of their instructors. The latter may seek consolidation of stall recoveries, while the former may be out to enjoy the luxury of the command in comfortable straight and level flight and a chance to have a relaxed look over the training area.

How do instructors get their students to adopt the same plan and thus the same precoding priorities? That's probably the most important question of all in teaching. One answer gets to the nub of the problem of motivating students. Basically the problem of motivation is *to get the student to do what the instructor regards as important.* Communication is an important factor.

Here are some good examples of communication difficulties taken from the *NAFI Foundation Newsletter* (July/Aug. 1984, 1):

> A flight instructor was directing a new student through his first approach to landing. Noticing that the airspeed was getting a little slow he told the student to "get the nose down." The instructor, maintaining a constant monologue while watching outside the cockpit, again noticed the pitch attitude was slightly high. Again he said, "Get your nose down." This time there was no response to his direction, so he repeated "Get your nose down!" At this point, the student finally replied "My nose *is* down!" The instructor looked at the student only to find him flying with his chin to his chest—*his* nose *was* down.
>
> After a student completed a touch and go in a *Citabria,* the tower called traffic at the 1 o'clock position. The instructor said, "I got it," meaning "I see the traffic." The student relaxed and let go of the stick, interpreting "I got it" to mean the instructor was now flying the aircraft. The instructor kept an eye on the outside traffic as the student watched the airspeed drop off to a stall buffet. The instructor recovered the airplane.
>
> While practicing a touch and go landing for a complex aircraft checkout, the student made a less-than-elegant landing. The instructor sensed that the student was rather tense and attempted to ease his mind by saying "CHEER UP, that landing wasn't so bad!" The student, hearing only what he expected to hear, interpreted the words as "GEAR UP . . . " The instructor luckily intercepted the student's hand on the way to the gear handle.

Directing Attention

We know, then, that the planning and precoding priorities of flight instructors and their students can be vastly different. But both instructor and student need to see the same task as important. The perceived relevance of the preflight brief or period of solo practice of stall recoveries is a contributing factor. An advance organizer, such as a clear statement of the objectives of the unit or flight, will certainly act as a communication. There is no reason why the student should not have a knowledge of the syllabus and course structure. Indeed, the student could be invited to draft a suggested program.

Perceiving the Task As Important

Part of the battle is won if the instructor can convince the learner that the task being undertaken is important or relevant. When learners choose

their own goals, there is no problem. There is a problem, however, when someone else does the choosing, because then the learners have to be convinced that the task is important.

This problem relates to precoding. A vital aspect of teaching is selling the task to the learner. The learner's scale of importance is the key. Good conversationalists recognize the importance of tapping the listener's interests: but the instructor's task goes beyond this. The instructor aims not so much at captivating the audience but at getting the learner to precode what is important and what is not. The less the learner knows about the lesson content, the more that directive "importance pointing" will be necessary.

Instructional Strategies

After the learner has agreed to engage the task, attention can be maintained by influencing his or her mental set to learn, immediately prior to, and in the course of, the learning itself. There are six strategies of directing attention in this way:

1. Pretest. Before a lesson, a set of questions related to the content of the lesson is asked. Normally, such pretests are used to determine how much students know already, so that when they have completed the learning episode and are tested afterwards (with a posttest), the difference gives an indication of how much they have learned. Such a procedure has an additional effect of alerting students to relevant material as they come across it.

This technique works for relatively short teaching episodes, and for either bright or mature students. It is also assumed, of course, that the learners already have some prior knowledge of the content, otherwise they wouldn't be able to make much sense of the questions. It should be noted that pretests work best for material that is relevant to the questions being asked: other material, not directly related to any question, is usually learned worse than if no questions had been asked. This finding certainly suggests that the effect is due to directing attention. If the aim is in fact to concentrate on particular points, then pretests are good. However, if the students are reading to explore the content, pretesting might actually inhibit them from seeking a fresh point of view.

2. Behavioral objectives. Behavioral objectives, detailing the student behavior expected at the end of the lessons, are an important means of directing attention. Their importance is such that we will return to them later.

3. Overviews. Overviews have been used for a very long time by teachers. These are simple statements of the main points to be dealt with, rather like abstracts or synopses at the beginning of the lesson or paper. They put in abbreviated form points and issues that will be elaborated on in the material immediately to follow. Again, the evidence is that overviews do

enhance learning, but the evidence is not so clear as it is in the cases of objectives and pretests. It is probable that overviews work, not so much through priming the attention of learners, as through the mechanisms of repetition and familiarization. That is, because the overview is written in much the same level of abstraction as the material itself, the student encounters the important concepts twice over: the student gets a double lesson, in effect.

4. Advance organizers. These are overviews of a special kind. An overview is written at the same level as the main material, whereas an advance organizer is a short passage that gives an abstract theoretical framework to interpret the material that follows. An advance organizer provides the concepts that will be used by the learner to understand the material to be learned.

Advance organizers help when the material to be learned is new and complex, and this applies to a large proportion of flight instruction.

5. Inserted questions. During a lesson or reading material, it is possible to extend pretest questions into the material itself, so that periodically the learner is stopped and questioned during the course of learning. This, too, seems to be an effective procedure.

6. Sample items. Students can be given items that typify those they will encounter in a posttest. These items can serve as a means of familiarization and for clarifying expectations.

Summary

Memory operates at three levels. In the sensory register, material is precoded very quickly according to its current importance to the individual. Importance is "decided" according to various sets of priorities, determined by: (1) the ongoing plan of the individuals, which is more or less a conscious choice of what they want to do; (2) the physical attributes of the stimulus, its intensity or variability; (3) the internal physiological states signaling biological importance.

Individuals vary greatly in what they consider to be important. Consequently their plans differ, and a frequent result is a communication gap. One of the major problems in flight instruction is bridging the communication gap between instructor and student. To complicate matters, priorities of importance often vary from moment to moment within the same individual, resulting in fluctuations of attention.

WORKING MEMORY

We now turn to the second box in Fig. 2.1, the working memory. Consider the following mental arithmetic problem: 333 multiplied by itself.

On the surface it may appear hard but it shouldn't be. One only has to know that $3 \times 3 = 9$, and how to add up. A nine year old should be able to do it easily, yet most adults find it difficult. Clearly this is not because it demands special knowledge, but rather because we need to hold the various steps in mind.

Normally, we solve the problem by writing down the progressive steps and calculations, in this way "externalizing" our memory. In mental arithmetic we need an internal space in which to spread our working. We call that space *working memory.*

Working memory is where conscious thought takes place: the dotted area in Fig. 2.1 represents the area, the contents of which we are conscious of at any given time. Working memory might be compared to a circle drawn on the ground one dark night with coins scattered on it. Now a flashlight beam is held at such a height that most of the beam is contained within the circle. The coins at the center of the circle will be seen clearly, those at the edge will be a bit indistinct, while those beyond the circle will be invisible.

Now say that the circle is such a size that only about seven coins can be contained within the circle. The currency we are using is information; the coins are what some psychologists call "chunks." If we use cents, then we shall only have seven units of information, but if we use nickels or quarters, we shall have very much more information for the same number of chunks. It is of course desirable to use informationally rich chunks where possible: *coding* is the process we use to convert our informational cents into dollars.

Some Properties of Working Memory

Working memory has three important properties:

1. *Working memory is limited in size.* We can attend to only one train of thought at a time and only to a fixed number of items. In adults that number is about seven. Our thinking is limited to a problem or situation that requires no more than seven elements or chunks of information. The 333×333 problem requires many more than seven chunks. And for an example, with words, if a sentence is too long and complex, we may fail to comprehend it. We may have forgotten the earlier part before we have come to the end.

2. *Working memory effectively increases with age.* Children have a much smaller working memory than adults. There is a steady increase in the number of chunks from early childhood to adolescence, after which it remains stable to old age.

3. *The two main processes deployed in working memory are rehearsal and coding.* If a chunk or series of chunks is neither coded nor rehearsed, it will not be placed into long-term memory and will not therefore be retrievable in the future.

The following numerical problem will illustrate this. Glance at the first row of digits below, and read from left to right at the rate of about one digit per second, saying the digit to yourself. Then close your eyes and write the series on a scrap of paper.

4 5 7 0 9

Now repeat the performance for the following series, dealing with each series at a time:

(a) 6 4 0 3 9 5 1
(b) 4 2 1 7 3 9 6 0
(c) 5 8 3 0 1 7 9 2 6
(d) 1 3 5 7 2 4 6 8 1 3 5 7 2 4 6 8

Which series was the easiest? Which was the hardest? Most people find (c) the hardest and (d) the easiest. On the surface that may seem odd, if we consider that (c) has only 9 digits whereas (d) has 16.

Of course the point is that there are not 16 "things to remember" in (d): there are 4. Odd numbers, even numbers, 8 the largest number, repeat. The series was coded into a more economical form using previous knowledge as the basis for the code.

How do people remember the other series (if they don't correspond to a known telephone number or some other relevant number)? They can just let the mind go blank and "hold," without doing anything active. This method is suitable if it is only necessary to hold the material for no more than thirty seconds, but if anything else distracts them, the material will be easily dislodged from working memory and cannot be retrieved from long-term memory. This is a familiar story when we look up a phone number in the directory, when we momentarily pause and think of something else; and then find we have to look up the number again.

Alternatively, one can keep the number by repeating it—recycling it—until dialing is finished. Repetition, or *rehearsal* as it is technically called, will help fix the number in long-term memory. If it can also be coded, in whole or in part, then so much the better.

Rote and Meaningful Learning

There are thus two kinds of processing that may occur in working memory to ensure that the material is learned for later recall: rehearsing and coding. Coding may be used where the material has some kind of structure, and the individual has the relevant background knowledge to make use of that structure. Obviously, someone who didn't know the difference between odd and even numbers would be unable to make use of

that code in remembering the series. For example, the pilot undertaking an endorsement on a new aircraft could utilize a personally significant code to remember the critical speeds. These speeds could be related to the equivalent speeds on another aircraft, to telephone numbers, birthdates, or some other numerals that are already in the long-term memory.

Rehearsal has the same compressing, or chunking, function as coding. It is used in physical skills and in verbal tasks where there is no intrinsic structure to the material or where the individual is unable to use what structure there is. Rehearsal is also used when the learner wants to make sure that learning is verbatim, or 100 percent accurate. Rehearsal is carried out on the actual words used, without reference to their meaning, and is called *rote learning*. Flight crews, for example, rote learn checks and procedures not because they do not understand them, but simply to ensure accuracy.

Meaningful learning by coding is much more economical, more stable, and usually more enjoyable than rote learning. It is carried out on the meanings that the words signify, rather than on the words themselves. Material that has been learned by coding may be reproduced in a transformed version of the original: the meaning remains the same but the words may be different. That, in fact, constitutes a crucial test that distinguishes rote from meaningful learning: rote learned material can't be replayed in a greatly changed way (obviously one can use synonyms for individual words) whereas material that has been learned with understanding can be entirely rephrased, if necessary. Thus, if a trainee is unable to explain a principle or a regulation in different words from those used in the rule book, the probability is that it has not been understood.

Recoding

There is another, more important, aspect to intelligence than working memory capacity. It is, in fact, rare that things happen in exactly the same way as they did before. How, under these circumstances, can a new experience be coded in terms of previous experiences?

It all depends upon how much the new experience has in common with what has been coded from past experience. To some extent, it will be necessary to change existing codes, so that the new experiences can be handled, and so that little of the old experience is lost. Sometimes the person changes too much, treating each new experience in terms of its novelty rather than its sameness.

More frequently, the person changes too little, distorting and misunderstanding the new experience so that it is forced to fit into his or her existing codes. People who do this are described as rigid, as they don't gain from experience.

Each new experience, then, is matched to what is already known. It is unlikely that the match will be exact, so there will usually be varying degrees of mismatch. When the mismatch is optimal, growth will occur. There is a change in code structure, called *recoding,* to accommodate the novel elements of the new experience with the elements of the old. What happens when this optimal mismatch occurs, then, is that the new content is learned, and the learner's codes have changed in a more complex, more sophisticated, direction.

Examples of such mismatches include judgments made about flaring a 747, when the pilot position is so much higher. Or the judgments involved while being so much in front of the wheels and turning onto taxiways. A simpler example is deciphering heavily accented radio communications overseas. What you expect to hear is not what you hear. You have to recode to match the new sounds.

There is one final aspect of recoding. Recoding is directly linked to intrinsic motivation, that is, our interest and involvement in a task. A task that demands no recoding is boring, we have seen it all before. For example, the lesson is over. It's time to turn for home. It can either be "OK, take us home" (no recoding here), or, "Just follow me through on this steep turn. This is what we will practice in our next lesson. Think about the control factors involved."

If there is some mismatch so that recoding is required, the individual is challenged. He or she finds the situation interesting, even exciting. However, if the mismatch is too great, it becomes overwhelming and threatening, while if there is total mismatch the situation is simply incomprehensible. The degree of intrinsic motivation experienced by a student thus depends upon the match between current experiences and the knowledge gained from previous experience. It is possible, sometimes, for instructors to design experiences for students that create the appropriate degree of mismatch.

Consider a pilot with extensive experience in propeller-driven aircraft converting to jets. Recoding has to take place with reference to spool up, sink rate, deceleration rate, and much more, including the possibility of sensitivity to new aircraft attitudes for certain maneuvers. The slower response during the seconds taken to spin the jets from idle to maximum, the related manipulation of the aircraft to control an approach when spool up to bleed valve closure is required, and adjustment to the need for five or six miles in which to slow the aircraft are typical of the recodings needed.

It is not surprising that a typical jet-jet conversion (say 727 to 737) could involve a three-month intensive course incorporating engineering, cockpit systems, simulator flight, and some 50 hours on line and clearance checks. One distinct advantage held by the check and training captain over the flight instructor is the fact that intrinsic motivation is rarely a problem in the pilot undertaking a check flight.

Generic and Surface Codes

It follows from the model of memory that the more meaningful the material, the better it is retained; and the more meaningful it is, the greater the number of connections with previous knowledge. In other words, meaningfulness varies, particularly according to the elaborateness of coding. It is helpful to distinguish between two kinds of coding: *surface* and *generic.*

Surface codes are not deeply implicated with previous knowledge. Material that is coded in this way is narrow in its range of application or in its transferability to different fields. For content that is coded in this way to be remembered, rehearsal will often be required to augment the coding. Surface coding includes a high proportion of rote learning.

In order for aviation subjects to be taught with a focus on generic codes, rather than surface codes, students should be asked to provide their own examples and applications to new situations. Students can be asked to say in their own words what they think the text or instructor is communicating. What does fuel/air ratio mean? What does the mixture control do? At what fuel/air ratio do the highest cylinder temperatures occur?

In answering such questions in their own words, students have to reveal an understanding rather than mere knowledge. Knowledge, facts, and detail (such as being able to pronounce "stoichiometric") are important. In generic code learning, however, they are means to an end rather than the end itself. In surface code learning, the facts are the end point in themselves.

An interesting distinction between surface and generic coding was made in an emergency exercise in which an experienced pilot was confronted by a center-engine failure on a 727. The pilot intuitively reacted by superfluous use of rudder. This was the surface code. Engine out, apply rudder. Instead of a selective application of knowledge, it was a mechanical, automatic response. Naturally, there were cues available to indicate which engine was out. In contrast, consider the solution of a 727 captain whose problem was the loss of control surfaces on the horizontal stabilizer. The book and surface codes do not lead to a successful landing by manipulating the differential thrust of upper and lower engines: that's using the generic codes!

The content to be taught should be broad in application and firmly based in what the learner already knows. The content should be relevant to the practical exercise, and what is expected of the student.

For example, Bernoulli's Theorem is technically a generic code of such fundamental importance that it explains flight itself. Yet students could see it simply as a collection of words to be recited on command. This type of learning is the result of only surface coding. In this situation, if a student is asked about Bernoulli's theorem a response may be: "Dynamic pressure plus static pressure equals a constant"; or, "When pressure energy is added to potential energy and dynamic energy, you get a constant."

And when asked to fly straight and level, the student goes through a series of actions in the correct sequence. There would be no link with Bernoulli's Theorem or its implications for stability or lift. This is a waste of both the instructor's and the student's time. What is needed is a generic coding in which the words describing the theorem are related to other knowledge and experience giving a special meaning to the actions of flying straight and level.

In other words, what one understands verbally is intimately related to what one decides to do. What one decides to do can be explained and justified verbally and at various levels of abstraction.

Surface coding leads to isolated learning; generic coding to integrated learning. Generic codes plug into a lot of what the student already knows, providing a new and powerful way of seeing things. Above all, they help the student put together a scheme for doing things: to fly with understanding.

Do Students Learn by Their Mistakes?

In an experiment, a researcher read his students an excerpt from a modern novel, and they wrote down what they remembered of it. Immediately after they had done this, the original version was reread. The following week, the students again wrote their version of the passage, and again the original was reread. This pattern, writing down and rereading, was repeated for four weekly test-relearn occasions. Now one might have expected under these conditions that the students would check their errors against each rereading. On the contrary, what the students reproduced each time was a version of their first reproduction, despite the fact that they had had four opportunities to correct their initial errors.

The implications of this study are quite important: instructors need to be very careful about what they say first time. If instructors find they were mistaken, and try to correct mistakes, they will find it difficult to get the second version across. Because of the likely high similarity between correct and incorrect versions, the second (correct) version is confused with the first (already learned) version.

Similarly, instructors who have given a test and then read out the correct answers may be disappointed. Logically, they might expect that the students who got their answers wrong will correct them, write the corrections in, and remember the corrections. Psychologically, however, it is more likely that the students will remember their original mistakes rather than the corrections, or simply become confused about what *was* the correct version.

The moral, of course, is to make sure students don't make mistakes in the first place.

Mnemonics

So far we have been talking about coding that makes use of a structure that is intrinsic to the content. We know that rehearsal is applied to material that has no evident or usable structure, such as rote learning numbers, dates, formulas, and so forth. But can coding be used to help memorize such material? It certainly can, and its use in such situations is achieved by *mnemonics.* A mnemonic is a symbolic device or cue for remembering pieces of information. A mnemonics system imposes a structure where none exists, thus conferring the benefits of ease of learning, firm lodging in long-term memory, and ease of retrieval, which are normally the prerogative of meaningful learning.

A very simple kind of mnemonic makes a word out of the first letters of the target words. The prestall checklist becomes A SEAL (Altitude, Security check, Engine check, Area, Lookout) or the trouble check before an emergency procedure becomes FMOST (Fuel, Mixture, Oil, Switches, Throttle).

On being introduced to prestart checklists, the student pilot can be encouraged to link first letters or activities with someone or something known.

For example

—Preflight check completed

—Seat adjusted, seat belt on

—Brake pressure tested, park brake on

can become PSB, or Pilots' Sexual Behavior, which should be unforgettable!

More elaborate mnemonic systems are more general. Typically, easily visualized objects are attached by sheer rote learning to a number or letter, and the function of memory is to retain the mixture of visual images rather than the original numbers. Fig. 2.2 shows a selection of memory aids provided by the U.S. National Association of Flight Instructors.

The most useful mnemonics are the simple acrostics such as A SEAL. Whenever a string of actions needs to be learned with 100 percent accuracy, it will be well worth the instructor's while to assist the student by providing a good mnemonic. The mnemonic in turn provides a helpful working structure.

Teaching Space-saving Strategies

Working memory space is fixed, problem solving requires working memory space, and more sophisticated problems require more space than simpler ones. The way to cope with sophisticated problems is by a great deal of rehearsal, thereby chunking the material so that it needs less space. This may then leave surplus working memory in which more aspects of the problem can be handled.

An example is flying a cross-wind approach. The beginner has so many

FIG. 2.2. **Mnemonics and Memory Aids in Flying**

These mnemonics are intended to serve as examples of the method used as a memory aid. The actual content may not apply in geographic areas according to their hemispheric location.

TRUE – MAGNETIC – COMPASS

Background: Three different reference points may be used to measure headings and winds. These reference points are: true north, magnetic north, and your compass. It follows that courses measured by reference to true north are the true course. A course measured in reference to magnetic north is the magnetic course, and after adjusting for compass deviation, it becomes a compass course or compass heading if it has been corrected for wind drift. The correction used to convert between true measurements and magnetic is called *variation,* which is found on your aeronautical chart.

The correction used to convert between magnetic measurements and compass measurements is called *deviation.* This should be found on a compass correction card mounted on the compass you would be using in flight. The following method is normally used to make the conversions.

	True Measurement
plus or minus	**V**ariation will yield a . . .
	Magnetic Measurement
plus or minus	**D**eviation will yield a . . .
	Compass Measurement

All of these memory aids are used to help remember the order of corrections. In other words, they represent a way of remembering TVMDC.

	Read This Way		*Read This Way*	
True	**T**(TEE)	**Turns**	**True**	**Twice**
±**Variation**	**V**(VEE)	**Vertical**	**Virgins**	**Vote**
Magnetic	**Makes**	**Make**	**Make**	**Men**
±**Deviation**	**Dull**	**Ducks**	**Dull**	**Dead**
Compass	**Children**	**Can**	**Companions**	**Can**
		Read This Way		*Read This Way*

"V" SPEEDS

V_y = best rate of climb
V_x = best angle of climb—"x" has more angles than "y"
V_{so} = Stalling speed in landing configuration—stalling with the **S**tuff **O**ut.

LATITUDE AND LONGITUDE

Background: Lines of longitude run between the North and South Poles, converging on the poles. Lines of latitude are concentric and run parallel to the equator. A few memory aids help keep the two straight.

Memory aids: Latitude is like a ladder—they remain parallel. **Long**itude lines are all **long.**

MORE ON COURSES & HEADINGS

Background: Any course corrected for wind will yield a heading. Hence, a true course ± wind corrections = true heading and magnetic course ± wind corrections = magnetic heading. Keep in mind the wind correction angle must be computed using a wind direction measured to the same reference as the course (i.e., true course needs true wind to yield true heading). Once you know true heading, you can use the TVMDC method to compute the compass heading. The following phrase can be used to remember the whole process of computing from a true course compass heading.

True course	+ Wind correction angle	= Heading (true)	± Variation	= Magnetic (heading)	± Deviation	= Compass (heading)
True	Wizards	Have	Very	Mean	Dumb	Cats

MAGNETIC COMPASS ERRORS

NORTHERLY TURNING ERROR

Background: While turning through northern headings, the compass indication will lag behind the actual heading of the aircraft. While turning through southern headings, the compass indication will lead the actual heading of the aircraft. The errors are maximum on headings of north and south and do not exist on headings of east and west.

Memory aid: THE SOUTH WILL LEAD AGAIN.

In the Southern Hemisphere, the mnemonic ONUS (Over North, Under South) is commonly used.

ACCELERATION ERROR

ANDS

Background: The acceleration error is maximum on headings of east and west. When you accelerate, the compass will indicate you are more north; when you decelerate, the compass will indicate you are more south.

Memory aid: ANDS (**A**ccelerate **N**orth **D**ecelerate **S**outh).

ALTIMETER ERRORS

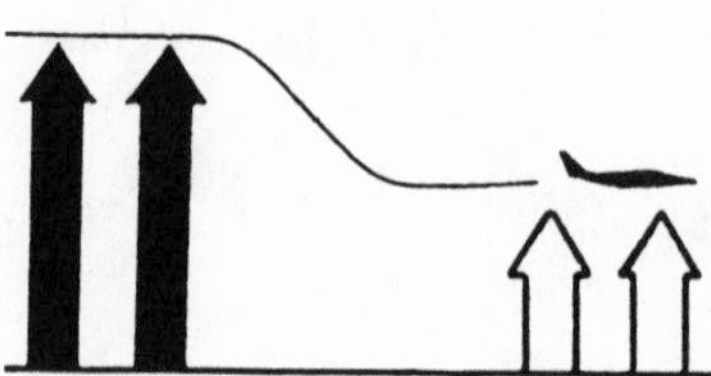

Background: When flying from an area of high temperature and/or pressure to an area of lower temperature and/or pressure, you will be lower than your altimeter indicates.

Memory aid: HIGH TO LOW, LOOK OUT BELOW.

COLD, WARM, STATIONARY, AND OCCLUDED FRONTS

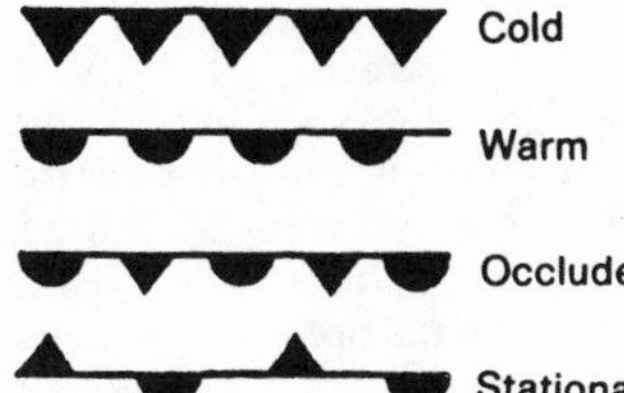

Cold front symbol shown in blue (it's cold!).
Icicles point the direction the front is moving.
Warm front symbol shown in red (it's hot!).
Blisters point in the direction that the front is moving.
Occluded front shown in purple (red + blue = purple) Cold front overtook warm front, blister and icicles show direction of movement.

Stationary front icicles and blisters are pulling in different directions. It's not going anywhere.

INTENSITY OF FRONTS

Background: The intensity of frontal weather from intense to less intense is as follows: **C**old, **O**ccluded, **W**arm, **S**tationary.

Memory aid: The first letter of each, in proper order of intensity, spells **COWS.**

DOWNWIND – UPWIND

Background: Students often have trouble differentiating between upwind and downwind.

Memory aid: Downwind is the same as downstream. Upwind is the same as upstream.

MARKER BEACONS

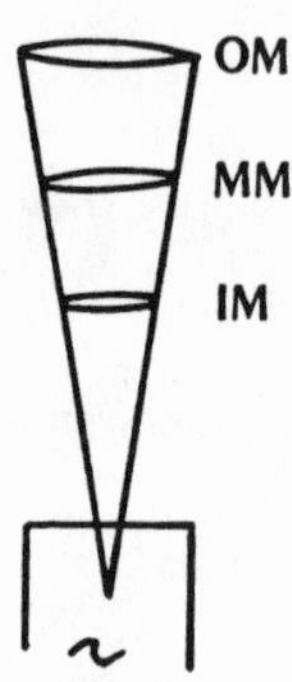

Background: Three different marker beacons may typically be found on an ILS front course approach. The outer marker (OM) is identified by a purple (or blue) light on the panel and an aural tone of continuous dashes at the rate of two dashes per second. The middle marker (MM) is identified by an amber light and an aural tone of alternate dots and dashes keyed at the rate of 95 dot/dash combinations per minute. The inner mark is identified by a white light and an aural tone of six dots per second.

Memory aid: The first letter of the colors spells out the word **PAW** for the OM, MM, and IM respectively. The aural signals may be remembered by comparing them to the heartbeat of a newly rated instrument pilot on his or her first solo ILS to minimums. OM – MM – IM.

VASI LIGHTS

Background: Two-bar VASI light system.

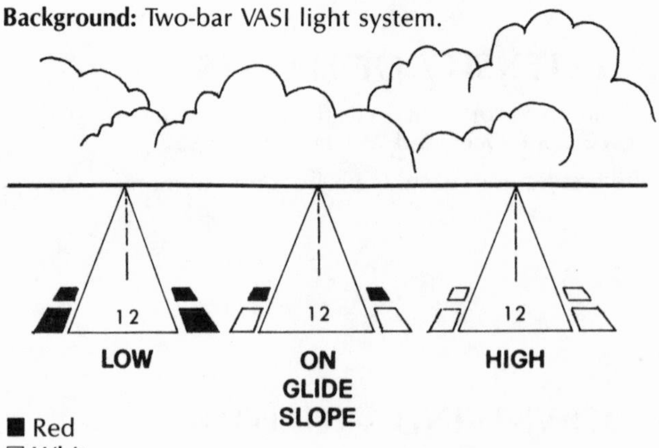

■ Red
□ White

Memory aid: Red over red you're dead. Red over white you're all right. White over white you're high as a kite.

Source: *NAFI Newsletter*, May/June 1984.

things to think about at once: power, attitude, drift, differential round-out, and so on. Under these circumstances the student might easily forget something crucial. The experienced pilot has practiced these things for so long that they are automatic: working memory space can easily be reserved for such things as airmanship, even for such irrelevant activities as maintaining a conversation.

One solution to the lack of working memory space is to analyze the problem and isolate those elements in the problem that make greatest demands on memory, and then drill the students in those demanding elements until they become automatic. Using this technique, students can solve problems that would, under conventional instruction, prove too difficult for them. Many other tasks can be analyzed into their "high load" components, and steps can be taken to reduce the working memory load they require. The use of crutches in mechanical arithmetic is an example. By writing a tiny 1 in the tens column, the student will remember to include the 10 that is to be carried from the units column. Experienced students who have automatized their procedures are likely to carry the figure without thinking about it, but beginners certainly will not. It is pointless to insist that working memory be cluttered with material that can more efficiently be "remembered" on a piece of paper.

Working memory is particularly likely to be crowded when the individual is in a stressful situation, such as in a sudden crisis in the air. Consider two scenarios: in what was planned as a simple stall recovery with power, a wing drops quickly; or on first solo, turbulence is encountered on late finals and the runway is obstructed.

What can be done to help the student facing potential overload?

The key is to cut down wherever possible on working memory load. This is not "spoon-feeding"; it is making sure that our limited minds are given the opportunity of handling more complex problems than they otherwise would.

The learner's strategy is to have automatic (or chunked) responses to key situations such as go-rounds. Instructors can ensure such automatic responses of grouped procedures by repeated practice in prior lessons. This is already common for procedures such as loss of power after takeoff.

In flight, the instructor's experience and learning usually provides reserve capacity to cope with further sensory input and its processing. The student, however, is probably working to capacity throughout the flight. This insight gives both an alternate criterion and an alternate structure for flight instruction. The job of the instructor now is to manipulate inputs so that the student progressively masters bigger "chunks."

The matter of capacity to process information is not just one of concern for *ab initio* pilots, however. There is a need to set up an airliner in the "keyhole" some 10 miles out from the airport. At 30 miles out and at 10,000 feet the jet will still be covering some 6 miles a minute at a time when the flight crew is experiencing its highest work load. The speed cannot be reduced; the rate of task accomplishment has to be accelerated. In such conditions, the pilot's task becomes one of ordering priorities. Consider the scenario: at 30 miles out it is possible that the crew could be required to undertake a frequency change, a course change, an aid change from Omni to ILS, a monitoring of the descent profile, radio calls, checklist items, and manipulation of advanced technology controls. And this is at 6 miles a minute. It is no wonder that the aircraft cockpit has been termed a "hostile environment" for teaching. In contrast, the briefing room is a haven from external distractions. The check and training needs of RPT pilots could be seen as falling within one of two broad areas of learning: first, manipulation of the aircraft, and second, operation of automatic equipment. The two overlap, of course, with the first being a prerequisite of the second.

While airlines provide manuals to guide flight crew and could be described as being procedurally oriented, it is obvious that the book cannot cover every eventuality. In check and training situations it is common for pilots in command to be able to react automatically (following the appropriate procedures), to an emergency (such as an engine out). It is when more than one emergency is provided at the same time that command judgment has to be exercised. Not surprisingly, it is such an eventuality that is most testing for flight crew. A fundamentalist and literal interpretation of a procedures manual is not the professional role of a pilot in command. Because it is impossible for any manual to completely cover every possible eventuality and all possible combinations of circumstances, a pilot in command will frequently have to make a decision to suit the situation that has arisen.

Working Memory Capacity and Intelligence

Given the utopian opportunity to fill a suitcase with coins from a bank vault, we are faced with a situation similar to that of the link between working memory capacity and intelligence. Consider the possibilities. A small suitcase (limited capacity) holds only a few coins. Used intelligently, the small suitcase could be filled with one dollar coins and perhaps have greater wealth than the larger suitcase unintelligently used to contain one cent coins: larger capacity, but less intelligence. Working memory capacity is a precondition for intelligent behavior, but does not guarantee it.

Summary

Working memory corresponds to span of consciousness. It may be regarded as an area in which chunks—items or groups of items of information—are held while we think about them. If the information is not processed by coding or rehearsal within about a minute, the information fades and cannot be recalled again. The number of chunks that can be held in working memory at any given time increases with age and is around seven or eight in adults.

Rehearsal is the process used in rote memory, coding the process in meaningful memory. In rote memory, it is difficult to recall the material in any order, or in a different structure and wording to that of the material originally learned. In meaningful learning, on the other hand, learners are able to explain the content in their own words and in their own ways.

Recoding takes place when learners adjust to new experiences and are enriched by them. Generic codes are those that have most meaning to students, and involve useful recoding.

LONG-TERM MEMORY

We now turn to what is usually meant by *memorizing*—learning material for its subsequent recall. A lending library offers a helpful analogy to this process. Books are usually cataloged on arrival by author and subject matter, and placed on the shelves accordingly. But there are other methods, of course. The librarian could catalog by publisher and year of publication or by the color and length of the spine. If the latter method were used, one could have a very tasteful arrangement with waves of color undulating around the room—but it might be rather difficult to retrieve any particular title.

The librarian's cataloging system corresponds to a coding system. Processing the books by rehearsal would correspond to a method that ignores the subject matter of the book. One difficulty of this analogy, however, is that it suggests that remembering is a retrieval process, like finding the catalog number of the book and retrieving the book from the stacks. Although many psychologists use the word "retrieval" for remembering, it is misleading. Retrieval implies that each memory is stored as a unique event in a particular location in the cerebral cortex. While it is certainly true that the general class of visual information tends to be stored in one part of the right hemisphere of the cortex, verbal in the left, and temporal and logical towards the front, memories within a sensory modality do not appear to have a specific location. For example, if some part of the brain is injured, a memory impairment (called an "amnesia" where loss of memory is involved) may occur, but in time, surrounding parts of the brain learn to take over and "cover" for the injured part, and a version of the event may be recalled.

This last example suggests that memory storage is more a matter of reconstruction than retrieval. What we retain and learn are a few specific details, plus a code or program that will reconstruct the original experience from those details. Consider how paleontologists reconstruct using existing knowledge (codes) a complete dinosaur from a few bone chips (specific details). To do this, they use the bone chips as a basis. Then, using their knowledge of anatomy and any other relevant knowledge, they make their best estimate of how the whole animal was constructed. They might be wrong in detail but generally correct in overall outline, which is much the same way as our memories serve us.

If recoding has taken place since the original experience, the codes to "unlock" that experience will have undergone change. In that event, recall of that experience is going to change. Recall of childhood experience in adulthood, for example, is usually inaccurate. It can be quite disillusioning to visit a place, well known and loved as a child, that hasn't been seen again for many years. The reconstruction and the immediate input are in conflict, often to the point where mismatch produces distress, not the comfortable glow one might have expected.

Dismembering

According to one theory of memory, an experience is *dis*membered into several components, which are then stored and synthesized when we *re*member. Dismembering may proceed along several dimensions.

First, there are the specific sensations, raw and uncoded. These are the bone chips. The ability to store such detail is very much stronger in children than in adults, and similar to eidetic imagery or photographic memory. These picture strips are mostly based on visual images, but they may occasionally be based on smell, taste, or sound.

Apart from the picture strip, then, there are four main dimensions along which input may be dismembered:

1. Semantic. This is the most common dimension, which refers to word meaning encoded as either verbal or auditory. We remember the general meanings of words comparatively easily unlike smell and tastes, for example, the storage of which is not very stable. The elaborate terminology that goes with wine tasting is an attempt to code an unstable experience in the stable structure of words. Semantic coding is of paramount importance in flight instruction, especially in checks and procedures.

2. Temporal. It is easy to remember things in the order in which they happened. If I want to remember what I did last Friday at 5:00 p.m., I can usually recall it by starting at some useful reference that I can remember (such as lunch), and then trace through the events following thereafter. Interestingly, the same strategy works for events that do not even form part of one's own personal experience. Aviation historians may find it easier to encode and recall aircraft types in terms of sheer succession, rather than in a more logical scheme, such as grouping according to payload or propulsion.

3. Spatial. Some information is better encoded in a spatial form, rather than semantically or temporally. For instance, in navigation, a map or a sketch diagram shows how objects are related in space and is far more economical and easier for most people to use than a series of verbal instructions. Visual spatial encoding can also be imposed on material that isn't inherently visual, for example, Fig. 2.1, where memory systems are represented in terms of a flow diagram. Many systems for improving memory rely on tagging the material with a visual image.

4. Logical. Memory is helped considerably if the material is encoded in terms of an existing structure, which has a logic of its own. It is much harder to remember illogical arguments than logically consistent ones.

The general relationship between dismembering and remembering is represented in Fig. 2.3.

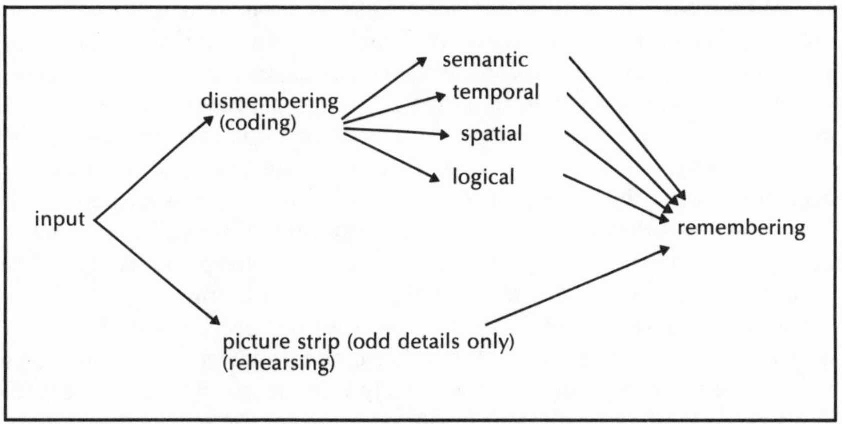

FIG. 2.3. **Dismembering and Remembering**

However, it must not be thought that these dimensions are mutually exclusive. On the contrary, as in a library, the more cross classification the more easily will the material be recovered. Just plain talk uses only the semantic dimensions. Chalk-and-talk is a bare improvement on this, depending on the instructor's blackboard skill. Remember, too, the point made about precoding; *attention is more easily held when there is variation in input.* Thus, good instruction should involve all possible means of encoding: striking images and diagrams to give a spatial encoding, use of mnemonic devices, presenting material in a logical step-by-step structure. Finally, sheer repetition should not be forgotten, especially for those raw picture strips of detail.

The overall process that this reconstructive model invokes is activity. Activity implies the active construction of meaning for stimuli, using verbal, visual imagery, and another sort of multiple processing. The more aspects of previous experience that can be linked to an item, and the more ways in which that item can be coded, the better will it be remembered. Activity is more important in learning then, because it helps link up the content learned with previous knowledge. It is especially helpful if activity provides cross-links with different modes of encoding the material.

Some Factors in Forgetting

There are two sets of factors in forgetting. The first is related to the reconstructive nature of learning, the second to interference when items learned are similar to each other.

Reconstruction

We have noted that our memories, like the paleontologist's dinosaur, may be inaccurate as to particular detail. The dinosaur is an educated guess, given the state of present knowledge. Thus, too, do our expectations often determine our memories. When groups of people reconstruct a past event witnessed by each person, the reconstructions are not always the same. Witnesses in law courts often swear in all sincerity that they saw the accused do what they *expected* to see him do, not always what he *did* do. One common form of change is called "conventionalization": one remembers an event as one might have thought it *ought* to have happened, rather than in the way in which it was actually reported as happening.

One of the writers was in the vicinity when a Piper Tomahawk crashed in a lake. Published accounts in the local press indicated that witnesses remembered that the aircraft immediately prior to striking the water: (1) was undertaking aerobatic maneuvers with power; (2) was undertaking a forced landing with intermittent power; (3) glided onto the lake surface; and (4) made a sharp turn at low altitude and hit the water with the port wing.

What seems to have happened here is that witnesses saw, or, to be more accurate, thought they saw, the aircraft's flight path immediately before impact. They superimposed their translation of their perceptions: a translation affected by their expectations.

Associative Interference

Learning new procedures may be inhibited if a previously learned procedure is similar. Teaching cross-wind landings is one point where flight instructors may encounter this associative interference at work. An accident prevention specialist for the FAA, Bob Heckman, regards cross-wind landings as one of the hardest things for the flight instructor to teach (*NAFI Newsletter,* June/July 1978):

> You know why? The people don't seem to know what to touch. They are still driving cars and turning the nose with the ailerons, instead of the rudder. And what happens to the cross-wind when the word comes up "variable"? Now, they drift right and left in a wild ballet. SOLUTION: teach slip approaches. Sure it's uncomfortable to hold that wing down, but to keep it straight he *must* hold rudder. If he holds rudder on final approach, he will hold on touchdown. He will learn to fly an airplane, not drive a car onto the runway.

When required to taxi an aircraft, car drivers inevitably use the aileron control to "drive" the aircraft around a corner in the taxiway. They have to unlearn this tendency, which is a previous learning that interferes with the behavior that is required.

After moving to another aircraft there may be an automatic reach for a trim control that isn't there in the different design.

There have been many experiments on the degree of similarity of stimulus, of response, and at what point after learning the confusing material occurs.

Such forgettings may occur very easily and thus no doubt play a large part in forgetting in everyday life, particularly where isolated or arbitrary units are used, such as foreign language words. Associative interference is less likely to occur in learning material that has been carefully learned and related to other knowledge. In fact, when we know a subject well, the more we read, the more differentiated and the more clearly do we sort out the details. It is in the early stages before we have built up this knowledge network, that associative interference, either retroactive or proactive, is most likely to cause confusion and difficulties.

Accordingly, then, instructors should be very careful in the presentation of new material in the early stages of a subject matter sequence. They should keep it plain and simple, and avoid detail, especially similar details that are easily confused with each other. Where there is a likelihood of confusing similar sounding names or events, attempts should be made to separate them as far as possible and fill in the interval with quite different material. Military instructors use the mnemonic KISS to guide their lesson plans: Keep It Simple, Stupid!

Summary and Conclusion

Experiences or events associated with flight are either coded or rehearsed (or both) when we commit them to memory. Procedures, for example, are dismembered into (1) specific detail that is stored "raw" on the basis of rehearsal; and (2) several dimensions. These dimensions include semantic meaning (where words or mnemonics are the code); temporal sequence (in which chronological order is the code); logical structure (in which the logic of the sequence is the code); and spatial imagery (in which the pilot works systematically from left to right or in a clockwise direction through instruments and switches).

Each of these dimensions has a different "home" in the brain. When the pilot remembers, the total memory is reconstructed by assembling all the components around the specific details.

Sources of error are conventionalization and interference. The former refers to the way we tend to remember things as they ought to have happened. The latter refers to the confusion of similar components in different contexts, as in lines of latitude and lines of longitude or variation and deviation. Instructors can provide learners with a variety of memory aids (exemplified in this chapter) to improve recall.

Learning and memorizing can be regarded as the selection and processing of information over various periods of time. Very broadly, two functions are distinguished: (1) selecting and maintaining attention to a particular theme, and (2) handling the selected material so that it is understood and available for future use.

For the instructor, the first problem is to direct the student's attention to task-relevant information. One general problem is that of a communication gap which can be minimized by utilizing student interest. This problem characterizes *ab initio* pilot training rather than commercial check and training operations, which are in the fortunate position of having the luxury of selecting from a large pool of flying talent. The selectivity factor virtually eliminates the communication gap in check and training operations. Nevertheless, highly structured strategies of maintaining attention still have application. These strategies include pretest questions, sample test items, overviews, behavioral objectives, advance organizers, and inserted questions. All can improve learning.

The second broad problem concerns the comprehension, assimilation, and recollection of information. Most flight instruction emphasizes meaningful (applicable) learning rather than rote (literal) learning. Generic coding is thus more appropriate than surface coding, which has few connections with existing knowledge. Testing for accuracy of recall rather than depth of understanding encourages the tendency for rote learning. Generic coding can be encouraged by instruction through use of examples and presentation of learning materials that cover a range of sensory inputs through, for example, overhead transparencies, videotapes, duplicated sheets, maps, models, cassette tapes, computer-assisted instruction, illustrations, and so on. The training film *The Wrong Stuff* shows how an airline crew used some original thinking to solve an electrical power problem by starting the APU. Such an understanding of aircraft systems, rather than mere recall of components, can be introduced into instruction as a "What would you do if . . .?" problem.

It was argued that students may not learn very easily from their mistakes. It is important, therefore, for the instructor to ensure that the material to be learned is correctly coded in the first instance, with the possible effects of associative interference minimized.

For the learner to retain the material in as efficient a form as possible, generic coding for maximum meaning is appropriate. In flight instruction, however, there will be times when literal accuracy is important (as in checks and procedures). In such instances, mnemonics may be of great value.

Training people to be better memorizers is a matter of helping them to be aware of effective strategies, such as rehearsing sufficiently, associating words and concepts with images, and using the structure of the content for generic coding. Coding and rehearsal strategies are attempts to beat working memory limitations. They also have the bonus of facilitating recall. A flight instructor needs to ensure that the student makes efficient use of working memory.

Discussion Questions

1. Nominate another exercise from a training syllabus. List in note form the major sensory inputs that would be registered by the student; the major information to be processed by the student; and, ideally, the information that would be stored in the student's long-term memory.
2. Your task is to *prevent* a student pilot from paying attention to you in (a) a preflight brief and (b) a flying exercise. In note form, briefly describe the methods you would use in each case.
3. For this exercise, name each of the six strategies that could be used for influencing a student's mental set, and provide one example of each.
4. Write one mnemonic that you use as a memory aid for students.
5. From your experience, cite one example of associative interference with flight education.

Further Reading

Biggs, J. B., and R. A. Telfer. *The Process of Learning*. Sydney: Prentice-Hall, 1987. Chapter 2 is entitled "Learning, Memory, and Activation." It incorporates details of six additional sources on: meaningful and rote learning; selective attention, coding, and rehearsal; recent memory research; mnemonics and long-term organization; the educational implications of memory processes; preinstructional strategies (including pretests, behavioral objectives, and advance organizers); and the educational implications of memory research.

3 Teaching the Skills of Flying

IMPARTING SKILLS is a central role of instructors in aviation. In this chapter the nature of a skill is examined in order to clarify the degrees to which a skill can be acquired and the conditions under which such an acquisition can occur.

The following questions are answered in this chapter: How do the cognitive, fixative, and autonomous levels of skill learning differ? How do central plans or "blueprints" help a student pilot learn to fly? Can an instructor provide feedback on performance in a way to help skill learning? Is it possible for a student to be a better pilot than the instructor?

After reading this chapter, you should be able to (1) identify three stages in the development of flying skills, (2) recognize the place of rehearsal and feedback in the learning of skills, (3) distinguish between intrinsic and extrinsic feedback to learners, (4) form a view on whether to teach the parts of a skill and piece them together or to rehearse the whole sequence, and (5) decide whether to distribute flight practice periods or whether to practice in massed periods.

The Nature of Skills

The student who experiences the standard stall recovery for the first time sees no intrinsic or logical reason why one part of the recovery procedure is related to another. Yet these activities have to be reproducible with speed and accuracy, and without conscious thought on the part of the pilot; that is, they coalesce to form a *skill.* The recovery procedure is an example of such a skill.

How can this occur? A clearer view can be obtained by examining the stages in the development of a skill.

STAGE 1: THE COGNITIVE LEVEL

At this stage the student is acquainted with the task and what is required. This is termed *cognitive* because the learner gets to know the task (but not necessarily at this stage how to perform it).

For example, the student can be told in a preflight brief that there are certain steps in stall recovery: control column forward; appropriate application of power; application of rudder to prevent further yaw; then when the aircraft is unstalled, level the wings with rudder and ease out of the descent.

STAGE 2: THE FIXATIVE LEVEL

Next learners follow through the procedure, at first slowly and laboriously, so that the sequences and movements are fixed in their minds. This stage is continued, as outlined in Chapters 2 and 4, largely through the application of practice, feedback, and instruction.

Andrew Detroi, accident prevention coordinator for the FAA, suggested two approaches to establishing the fixative level in the *NAFI Newsletter* (April 1978; June/July 1978).

> APPROACH #1:
>
> To get the proper habits formed, we must repeat maneuvers *correctly* many times and *avoid* doing them incorrectly. Allowing students to use different procedures at different times destroys the habit-building process. This often happens when a student alternates instructors. Chief instructors must make sure that all instruction in their school is thoroughly standardized.
>
> Let us examine the above in practical terms. Most instructors would state unhesitatingly that their students do maneuvers correctly. But do they? I fly with many pilots. Many do not look before they turn, and those that do often do so incorrectly. Have they forgotten since "graduation"? Not on your life. Had they ever developed the right habits, they would still be there. What normally happens is that instructors scold their students *after* having done something wrong instead of preventing an improper act. Some instructors eventually go on to more advanced work and quit enforcing the basics altogether so the habit never develops. What should be done is this: Describe the act in detail and set up a "by the numbers" procedure for it. (You must write it down and make the student learn it.) That is:
>
> 1. With wings level, look opposite the intended turn (we don't want to turn our tail to another craft that may be on a collision course).
> 2. Look toward the intended turn as close to 180° as possible. (Here is where you really do the "clearing," especially in a high-wing airplane.)

3. Start the turn *and hold a predetermined bank* while you keep looking outside the cockpit most of the time.

4. Lead your rollout and stop turn on a predetermined section line 90° from start.

Once this is understood, do not allow the student to complete the maneuver unless it is done exactly as you recommend it. If any step is omitted, take the controls, return the aircraft to its starting point and just sit still. The student will know why. In just a few subsequent trials the procedures will become habit and no more problems will occur. There may be no exceptions, however, and *all* maneuvers throughout the training must be done often enough to maintain the habit. If, for instance, we teach flight at minimum control speed just before solo and then not practice it until the student is ready for recommendation, the proper habits probably will not be formed. Because landings are practiced just before solo, there is no reason to neglect air work. In the beginning of each flight, your student may go through a sequence that strengthens his grip on flying tremendously. For instance: Takeoff—break pattern—climbing turns (either section to section or 45° either side of a landmark or line)—level off to cruise—flight at minimum control speed—departure stall—landing stall—accelerated stall—return to cruise on a section line—normal glide—flaps down in glide and back to no flap—then back to cruise configuration.

Encourage students to do more and more in sequence *on their own.* If, for instance, he couldn't perform the above sequence without help, he certainly is not ready to solo. Instructors often stay too close to their charges, guiding them and guarding them so closely that they just don't develop self-reliance. This also keeps the instructor from finding out just where the weaknesses are. Don't just keep talking incessantly. There comes a time when the student should be able to perform on his own. Try this: give your 7- to 15-hour student enough to do (in sequence) to keep him busy for 5 to 10 minutes. Now *sit on your hands and close your mouth* to see how he does. Some instructors I know will get an awful shock. If your student does not do as well when left to his own resources as you thought he would, then it is time to look at your own work with a more critical eye. If you do not have consistent procedures for teaching or flying, then how can you instill the same in someone else? We cannot practice something that is not well defined and organized in our own minds.

To summarize:

1. You must define each task and break it into steps your student can learn in order.

2. The student must demonstrate to you that he knows those steps on the ground.

3. The student must be made to perform exactly as briefed and

not allowed to continue a maneuver unless it is done in the order prescribed.

4. Try to do most maneuvers on *all* the rides to maintain currency.

5. Delegate more and more of the responsibility for the flight to the student, letting him perform chains of already known steps without any comment from you. This builds confidence, motivates him, and allows you to evaluate.

APPROACH #2:

As a general rule, instructors have long neglected the best tool they have, which is aircraft attitude. If I had to select one indispensable concept for teaching someone to fly, it would be that aircraft are controlled by attitude and power. Since normal climbs, cruise and glide use a predetermined power setting, you only have to teach students to recognize the appropriate attitudes and all will become much easier. The best instructors I have known make attitude the focal point of basic flying because if you maintain the appropriate attitude for a given maneuver everything else remains constant. If your students can't express what they are doing in terms of attitude changes, then they really don't understand how an aircraft behaves.

Perhaps the best way to introduce this idea to a new student is to draw a vertical grease pencil line on the windshield right in front of him. (Not the center of the windshield, except in tandem aircraft.) Draw horizontal cross-marks two inches apart and number them top to bottom. In level cruise, have the student determine which spot of his grid covers the horizon. Holding that attitude equals level flight, even as he rolls into turns. The elevator pressure necessary to keep the spot on the horizon will vary with bank but if the spot is kept there, you will hold altitude. (CAUTION: If airspeed is changed appreciably, a different spot will be on the horizon.) Density altitude, load and seat position also will vary the "spot" but that is all right too. What we are teaching is awareness of our attitude and the ability to hold it constant. Hold whatever you think is right (but hold it, damn it) until everything becomes steady state. Now if airspeed, vertical speed or altitude is not what you want, you can make a small change to bring about what you need.

Next, hold a normal climb and find the "spot" for it. Do the same for the glide. One should have to refer to the airspeed only occasionally and still make the aircraft perform as desired. Later, introduce other "spots" for full flap glide, 1.3V_{so} final, flight at minimum control speed, V_x, liftoff, etc. If your student cannot make a takeoff, assume climb attitude, trim for this attitude and *know* that his airspeed is right without looking at it, then you missed the most important key to teaching. If he can't explain how aircraft attitude is affected by flap and power changes he should not be flying solo because he does not understand how his aircraft behaves. Does your student really know how to control his aircraft or is he just being taken for a ride by the airplane he is in? Think it over, if any concept really makes a difference—this is it!

STAGE 3: THE AUTONOMOUS LEVEL

The autonomous level is reached when the skill runs off on its own. This is the desired end-state for skill learning. The "blueprint" for the skill has been learned and it runs off as required. At that point, a high level of skills has been attained. For example, the student will automatically recognize stalling in any configuration and will go on to just as automatically carry out the standard stall recovery.

By this stage, the skill itself is not thought about consciously. Instead of occupying virtually the whole of working memory capacity, the skill occupies very little, so that more important things can be thought about—such as an instrument approach or a heavy work load situation, such as determining an alternate because of a change from VMC to IMC.

Note that the student has not learned merely a sequence of movements. Certainly serial learning may be involved in the earlier fixative change (control column forward—power on—right rudder—level wings—climb).

The fixative stage involves what researchers have termed "closed loop control" in which each movement is at first independent of the other. The stall recovery movements become dependent in the sense that one movement (control column forward) became the signal for the following movement (power on). In other words, the chain of movements is closed in the sense that each movement leads to the next, which leads to the next, and so on. A series of movements is still involved, not yet a single act.

This is not what happens at the autonomous stage. The hallmark of the autonomous stage is precisely that these deliberate, one-to-one, verbal-motor associations may be dispensed with. The autonomous stage is characterized by "open loop control," which operates independently of external feedback from preceding movements. The professional football player does not learn a series of if-then movements rigidly tied to each other. In a flash he computes (but unconsciously) how he needs to adjust his total response to account for change in distance, speed of ball, wind speeds, what other players are doing, and so forth. This is really a phenomenally complex calculation that is obviously a higher order activity than rehearsing and putting together all the individual movements that can be made.

In short, skills are learned sequences of activity that (a) are run off rapidly and accurately; (b) involve minimal conscious effort; (c) are nonetheless highly complex, and require considerable work to reach the standards of (a) and (b); and (d) are controlled by central plans, not one-to-one verbal or motor associations.

Some Conditions of Skill Learning

Four conditions associated with the learning of skills are discussed in this section: rehearsal, feedback, whole versus part methods, and massed versus distributed practice.

REHEARSAL

How are skills learned? The answer comes from the discussion in Chapter 2 where it was pointed out that the transfer of material from the short-term to the long-term memory was via chunking. As far as skills are concerned, chunking is achieved by means of rehearsal. The immediate actions, say, in the event of fire in the air or of one engine failure in a twin engined aircraft, need to be rehearsed again and again until the whole sequence runs off automatically in the correct manner.

It is worth noting that rehearsal does not have to be carried out physically. Motor skills can be "rehearsed" mentally. Performance may be improved by imagining that one is carrying out the emergency procedure. It will save precious time if students are encouraged to imagine the skills they have to perform.

FEEDBACK

Rehearsal by itself is hardly sufficient to learn a skill. If you never knew what you were doing was right, then clearly you might be rehearsing and learning the wrong thing. Feedback, that is, the knowledge that what you are doing is correct or incorrect, and if incorrect in what way incorrect—is absolutely vital. Any skill-learning situation must be one that provides adequate feedback to the learner. The provision of feedback is the instructor's primary responsibility in skill learning. As far as flight training is concerned, the issue is to provide ongoing feedback to the student as accurately and as simultaneously as possible. It is usually difficult for the learner to see whether he or she is right or wrong. At least when learning meaningful material, learners get clear signals as to whether they are generally on the right track or not. When the comment is made to oneself, "I just don't understand this," it is at least a personal provision of feedback. While it may not tell one what to do next, one at least knows that present actions are incorrect or inadequate.

The skill learner is in a less happy situation. As army instructors have been known to bark, "You don't have to understand, Soldier. Just *do* it!"

And, as the army instructor also knows very well, the more the recruits do it, and the more feedback they get, the more efficiently they will learn the skill in question. Of course, the best form of feedback is not necessarily that which comes from an eagle-eyed drill sergeant who takes a grim delight in bellowing "WRONG!!" at every false move. But at least it's feedback.

Feedback is of two kinds, extrinsic and intrinsic. *Extrinsic feedback,* or knowledge of results, is provided from a source external to learners who see or hear the result as soon after the act as possible. The flight instructor is most effective when, instead of giving individuals a rundown of where they went wrong after they have completed the flight, he or she does so during the lesson. Video film taken of the learner is another means of providing feedback. The advantage here is that learners can see for themselves what they are doing wrong; they don't have to rely on the word of someone else, which is often difficult to take on trust.

Intrinsic feedback is independent of an outside agent and is provided to the learners on the basis of how their actions feel to them. This is the best kind of feedback to have because it is most convincing, and it dispenses with other agents or expensive instruments. The possibility of intrinsic feedback is, however, limited in many situations. Take such skills as hitting a golf ball, kicking a football, rifle shooting, and learning to type. In all these skills, learners can tell for themselves whether they have achieved the desired result, but that is about all. The intrinsic feedback from their movements—prior to the final one that leads to the desired result—is inadequate and does not easily tell them how the shot might have been improved. In other skills, such as diving or dancing, intrinsic feedback is even more difficult to interpret from the point of view of the final result. Intrinsic feedback in these cases is insufficient, particularly in the formative stages of learning.

Nevertheless, the ideal form of feedback is intrinsic, where learners can themselves, at any point, find out how well they are performing the act. Now this state is achieved in the case of highly experienced performers: an Olympic shot-putter knows before a throw has been measured whether it was a good one or not. Somewhere stamped inside the athlete, there is a perfect template or blueprint; and the particular act as it is run off matches the blueprint or does not. It feels like a good shot or a bad shot. The deviation from the blueprint is noted by the performer, and the performer doesn't need to be told by an outsider when or where the performance was not wholly correct.

This idea of internalizing an ideal standard puts the feedback issue somewhat differently. The problem is to get the blueprint "there," inside learners, so that they can construct a better and better program on the basis of the mismatch feedback they get from the blueprint. Experience is obviously one way of learning such a blueprint, although it is very slow. It also rather begs the question: the problem is not one of helping the experienced performer but of helping the *in*experienced performer to internalize a criterion of perfect performance.

This is where it is important to have a good model in the right-hand seat—the chances are that the student will find the related skill easier to internalize by modeling. This point is very well illustrated by a story concerning Charles Chaplin. In the middle of a party, Chaplin rendered a stirring excerpt from an Italian opera. A guest exclaimed "Why Charlie, I didn't know you could sing!" Chaplin replied: "I wasn't singing; I was only imitating Caruso!" The moral to the student: "Don't fly, imitate your instructor!"—up to a point, anyway.

The cognitive stage is a general and imprecise step towards internalizing a blueprint. The limitations of a pure cognitive or verbal understanding of the skill are obvious: it does not tell the learner what it *feels* like to carry out the act perfectly, or even what it feels like to carry it out imperfectly. It seems important, then, to expose the learner to good models. Periods of skills acquisition might be spent in watching skilled performers, on film or in the flesh. When students have had long exposure to such models, they might begin to grasp what a good performance feels like.

Modeling is obviously a secondhand way of coping: success in learning is marked by an acceptably close reproduction of the instructional objective. Can't learners eventually do better than the set standards, or better than their instructors?

There must be many top pilots whose first standards of "perfect" performance were set by a mediocre instructor. What that possibly not-so-hot instructor did was to help the learner take that most important step of internalizing some criterion to the point where the student could generate his or her own ongoing intrinsic feedback. Hence, the first solo. By that stage, the learner will have internalized enough to form some kind of blueprint, from which he or she could develop flying skills relatively autonomously.

However, there are those who feel too much importance has been given this teaching tool. A critical perspective on the "learning by watching" idea has been provided by Dr. Stan Roscoe, professor of psychology at New Mexico State University and authority on aviation psychology and human factors. He points out that repeated efforts to demonstrate the benefits of modeling in quantitative terms have yielded consistently inconclusive results. He goes on, "If one could learn anything helpful from watching skilled performance, I would be a good golfer, I might even win an Olympic gold in gymnastics. I think this idea is vastly overrated by flight instructors and that their repeated demonstrations of 'how to do it right' merely steal time from the student."

WHOLE VERSUS PART METHODS

One important issue in the teaching of skills is the whole versus part question. Should the learner be taught pieces of the skill and then fit them together, or should the whole sequence be rehearsed each time from beginning to end? This has been possibly the most researched issue in skill learning. The most common answer provided by such research is that the whole method is superior, but this answer must be heavily qualified.

In point of fact, it is not very helpful to conceive the issue in this way. It depends very much upon (a) the skill in question; (b) whether we are talking about the cognitive, fixative, or autonomous stages of skill learning; (c) what we mean by "whole method" and "part method" (there are many variations of each); and not least (d) what the working memory and attentional capacities of the learner are.

First, it is an oversimplification to divide a skill into wholes and parts. What is the unit that is the "whole" act? Solo cross-country, solo circuits, solo? Usually, one refers to the whole of the required task as the unit in question. Logically, one self-contained sequence of flight would constitute the whole.

Second, where do you define the starting point in part of a skill—the beginning of an act or the end? If you start at the beginning, and proceed forward in bits, you have what is called the *progressive-part method.* If on the other hand, you start at the end of the act and proceed backward to the beginning, you have the *reverse-part method.*

The whole-part question falls into place when we consider the three stages in skill learning. At the cognitive stage, it seems necessary to give learners an idea of the whole skill: not to do so is to prevent them from forming an adequate overall plan. At the fixative stage it may well be that the learner is well able to perform some parts of the skill but experiences difficulty with others. In that case it seems rather stupid to insist that he or she has to go through the whole lot each time, if it can be avoided; it is wasteful and boring. At the autonomous stage, on the other hand, we are again talking about the whole skill—and we would probably revert to practicing the whole sequence until the learner's plateau had been reached.

R. M. Gagne, speaking to the Division of Military Psychology, American Psychological Association in New York (5 Sept. 1961) said this about design of training:

> 1. Any human task may be analyzed into a set of component tasks which are quite distinct from each other in terms of the experimental operations needed to produce them.
>
> 2. These task components are mediators of the final task performance; that is, their presence insures positive transfer to a final performance, and their absence reduces such transfer to near zero.
>
> 3. The basic principles of training design consist of (a) identifying the component tasks of a final performance; (b) insuring that each of these component tasks is fully achieved; and (c) arranging the total learning situation in a sequence which will insure optimal mediational effects from one component to another.

In flying a normal powered flight pattern, the instructor demonstrates the whole exercise. Then he or she extracts the skill elements for instruction, then relinks them. For example, the takeoff skill breaks down to preflight checks, prestart checks, start, after-start checks, taxiing, run-up, pretakeoff checks, and takeoff. But in all this, the instructor has to bear in mind the capabilities of the learner. All aspects of a skill, especially in the early acquisition stages, demand a heavy working memory load. If the act is so long that the learner forgets critical information about linking one sequence with another, then the whole method will be very inefficient. In such a case, it might well be more effective to break up a complex act into parts, and then when this part has been chunked by rehearsal, there would be profit in linking up with new aspects.

The question of instruction with respect to whole or part strategies thus requires some commonsense analysis. Certainly it is possible for learners of roughly similar abilities (including working memory capacity), to find out whether a whole, reverse-part, or progressive-part method is best at the various stages of skill learning. At Williams Air Force base, reverse-part strategies proved more effective in instruction of air-to-surface weapon training. Flight instructors should do their own informal research on this and similar matters, simply by keeping systematic records of how students progress when taught by these teaching strategies.

MASSED VERSUS DISTRIBUTED PRACTICE

Another issue in skill learning is whether the teacher should distribute practice or give it in massed amounts. Is it, for example, more effective to practice flying skills for a lesson a day throughout the week or for four lessons a day over the weekend? This question closely resembles the previous one, in the sense that one cannot provide absolute answers. One learner may become fatigued much more quickly than another; a highly motivated learner can practice for hours on end before getting bored; very short durations of practice may be insufficient for the learner to warm up or to become involved in the task.

One important factor here is again the working memory capacity of the learner. If the time between practices is too long, the learner might have forgotten what had been learned the previous time, so that the learner in effect starts from the same baseline each session. Another factor is the nature of the task, particularly whether it is physically demanding.

Instead of looking for general answers, then, the instructor should take a close look at the nature of the skill and the demands it makes on the student. The major issue appears to be off-setting losses due to forgetting against losses due to fatigue. One view is that in flight instruction this balance changes with student progress. Initially, the problem will be forgetting rather than fatigue. One senior instructor found in his experience that it was ideal for the student to fly one lesson a day up to the solo stage. This produces an opportunity for the student to think things over, rehearse (including mental rehearsal), prepare, analyze, and discuss.

In a study by Mengelkoch, Adams, and Gainer on the forgetting of previously learned flying skills, researchers had two groups of pilots fly the mission shown in Fig. 3.1 on an aircraft simulator. After an introductory period, one group was given 5 training trials and the other was given 10 training trials of 50 minutes. Each trial was a structured series of maneuvers and procedures from starting the engine, to takeoff, and on to landing and engine shutdown. A retention interval of 4 months followed the last training trial for both groups. The principal result was that the greatest decay in skill, regardless of training time, was in procedures. Proficiency in controlling the flight parameters (as in tracking) was never operationally significant in its deterioration. The authors concluded, "The findings of this study strongly suggest that programs directed toward the maintenance of flying proficiency should focus on the training of procedures." The study also has implications for instruction in emergency procedures.

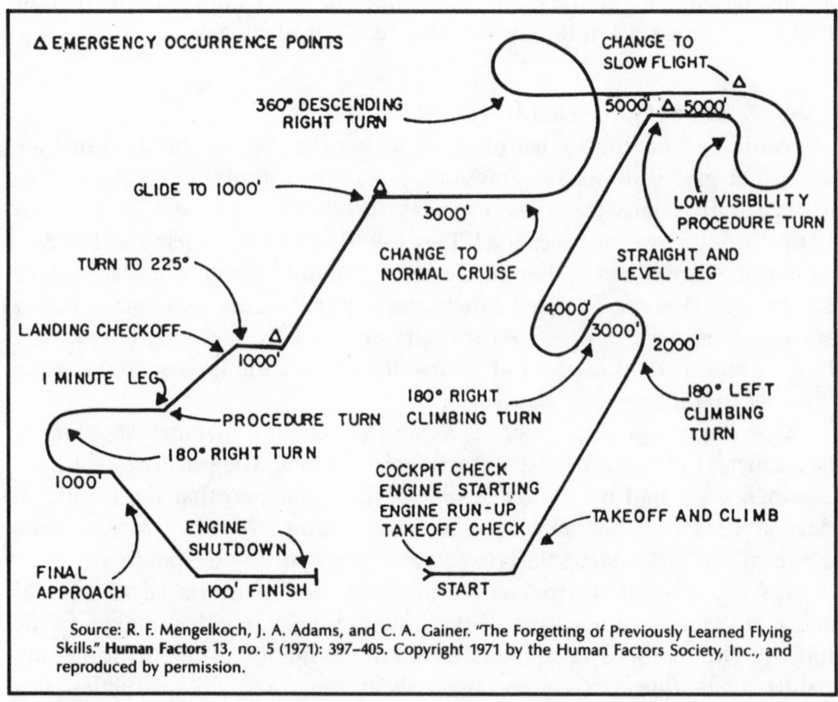

Source: R. F. Mengelkoch, J. A. Adams, and C. A. Gainer. "The Forgetting of Previously Learned Flying Skills." **Human Factors** 13, no. 5 (1971): 397–405. Copyright 1971 by the Human Factors Society, Inc., and reproduced by permission.

FIG. 3.1. **The Mission Sequence Flown to Study Decay in Flying Skills**

It is the instructor's estimate of the stage of skill acquisition and the extent of decay of procedural skill that will lead to the conclusion concerning massed versus distributed practice. In practice, for the *ab initio* student the actual determinant may be the combination of time and money constraints.

Summary and Conclusion

In this chapter we have been concerned with instruction with respect to the teaching of skills. Skill learning is where the contents of the learned act have no intrinsic relation to each other. We speak of skill learning (rather than concept or knowledge learning) when the material is learned to the point where it may be run off with speed, accuracy, and the minimum of conscious control.

Three stages of skill learning may however be recognized—cognitive, fixative and autonomous—and particular objectives would vary according to the stage currently being performed.

Two basic principles underlie instructional procedure in skill learning. First, acquisition is based upon rehearsal; and second, the learner requires feedback during and after rehearsal. Other conditions of learning—whole versus part, massed versus distributed—may be regarded as describing particular conditions under which rehearsal-with-feedback may take place. It was not, however, possible to prescribe general rules for the operation of massing or distributing rehearsal, or for rehearsing by the part or by the whole. These issues are too relative and too complex for general prescriptions to be of much use. Each instructor needs to work out what works best.

A sounder strategy for devising methods of instruction and training is open-ended. Given adequate rehearsal, one of the most critical factors governing skill acquisition is the provision of feedback—this is the area in which there is most slip and in which the instructor has the greatest potential for maximizing learning. The best form of feedback is intrinsic, which is generated and interpreted by learners themselves during the performance. Unfortunately, there is a direct relationship between efficiency in the skill and efficiency in utilizing intrinsic feedback: skilled performers already use intrinsic feedback. The problem is to encourage sensitivity to intrinsic feedback in the unskilled learner.

The instructor's task is therefore to optimize the conditions under which students set up internal criteria of what constitutes "good" performance, so that they are in a position to match their movements with the internalized ideal. In particular skills, it may be possible to use a kind of "imprinting," but in general the technique of providing plenty of good examples or models is recommended. Eventually, it would be hoped that the performer outgrows (and hopefully, too, outperforms) the initial models. Modeling involves much more than simply setting standards: there are likely to be strong affective and motivational components, and these may easily be capitalized upon.

In brief, if the instructor bears in mind the overriding importance of intrinsic feedback and works towards that with whatever devices a fertile mind can devise, he or she will do rather better than with heavy reliance upon sheer extrinsic feedback (which may carry the additional danger of being interpreted as punishment). Another most important aspect of feedback is that it should be as immediate as possible—as near to simultaneity with the act as possible.

Skill learning, then, is not just a simple matter of learning to do certain movements at a certain time. What is learned is a highly complex plan that allows considerable flexibility as to specific movements. While performing a skill at the autonomous stage does not involve conscious thought, the acquisition of skills properly requires a lot of thought by the student and perhaps especially by the instructor.

Discussion Questions

1. Choose another flying exercise from a syllabus, in this case making it one which has a clear orientation to the acquisition of a skill by the student. If more than one skill is involved, isolate one for this analysis. Then briefly outline the method you would use to facilitate the three stages of skill acquisition by the student: the cognitive level, the fixative level, and the autonomous level.
2. You are going to be asked to participate in a debate on the best method of teaching flying skills: Whole or part? Massed or distributive practice? Prepare, in brief note form, the advantages and disadvantages you see with each method. Use practical examples to reinforce your argument.

Further Reading

Curzon, L. B. *An Outline of the Principles and Practice: Teaching in Further Education.* 2d ed. London: Cassell, 1980. Chapter 14 on the skills lesson provides an informative overview of four other approaches to an understanding of how skills are acquired. Apart from the three-phase theory used as the preferred model in the present volume, Curzon describes the following: (a) selection theory, in which the learner is refining the process of selecting the most appropriate methods from his repertoire; (b) graded-movement theory, in which the learning of graded movements occurs in two stages. Stage 1 is a verbal-motor phase when the teacher provides verbal cues about the student's actions. Stage 2 is the motor phase in which the verbal cues are no longer necessary; (c) redundancy-appreciation theory suggests that the learner acquires skills as the student is able to understand the redundancy in inputs of sensory information; and (d) hierarchical-structure theory skills can be viewed as the progressive coordination of separate units of activity into a hierarchical structure.

Mengelkoch, R. F., J. A. Adams, and C. A. Gainer. "The Forgetting of Previously Learned Flying Skills." *Human Factors* 13, no. 5 (1971): 397–405.

Nance, John J. *Blind Trust—How Deregulation Has Jeopardized Airline Safety and What You Can Do About It.* New York: William Morrow, 1986. In conjunction with a theme of safety, airmanship, and the importance of judgment, this book offers both instructor and student a fascinating insight into the human factors in aviation.

Roscoe, S. N., et al. *Aviation Psychology.* Ames: Iowa State University Press, 1980. Chapter 1 on concepts and definitions is a very helpful starting point, especially the functional model of pilot-airplane system operations. Roscoe calls for research on differential rates of decay in flight skills and on efficient and economical methods of maintaining uniform levels of proficiency. Chapters 15–22 are on training, with an excellent introduction by Roscoe, Jensen, and Gawron who emphasize the centrality of the instructor's role as the greatest single source of variability in pilot training. Other foci are the measurement of the transfer of training, flight trainers, simulators, adaptive training, and computer-assisted flight training.

Motivation and Arousal: General Determinants of Behavior

THE FOLLOWING QUESTIONS are answered in this chapter: How does reaction to a "May Day" call relate to precoding? What is the optimal level of student arousal for preflight questioning by an instructor, or a complex flight maneuver? How does anxiety affect a pilot's performance? How can a pilot overcome the effects of stress? How can a student overcome test anxiety?

After reading this chapter, you should be able to (1) distinguish between the general drive of students to perform and the particular acts they are specifically motivated to perform, (2) link the arousal system with the sensory register and working memory, (3) show the difference between the energizing and interfering effects of arousal, (4) relate level of arousal to task performance, and (5) use variability in your instruction to create an optimal level of arousal in students.

Two Aspects of Motivation

Motivation has to do with feelings and emotions: the *affective domain* as it is called. The content of the previous two chapters on the other hand—learning, thinking, memorizing, and problem solving—is called the *cognitive domain.* Motivation may be considered on two levels: the general drive to behave and the particular acts that one is specifically motivated to perform. Under the general heading, consider how on certain days people feel quite literally "on a high," and eager to do things; while on other days they are down and find it quite difficult to do anything. Clearly, this phenomenon has a lot to do with motivation, and instructors should understand something about it. We shall look at this general level of motivation in this chapter and consider more specific aspects in the next.

The Concept of Arousal

The concept of arousal is the key to understanding the general level of motivation. Now Fig. 4.1 introduces a fourth box, labeled "arousal system," to the memory system described in Chapter 2.

The arousal system is linked with both the sensory register and working memory by paths 2 and 3 respectively. These have a crucial effect on mental processes, particularly from that first moment we pay attention to something.

Recall the discussion in Chapter 2 concerning precoding: when something is precoded as important, the message goes straight to working memory for further processing. This is marked as Pathway 1 in the diagram. In addition, however, another signal goes to the arousal system (Pathway 2). It is this boost to arousal that motivationally distinguishes important from unimportant messages. Finally, the arousal system affects working memory (Pathway 3) in the ways described below.

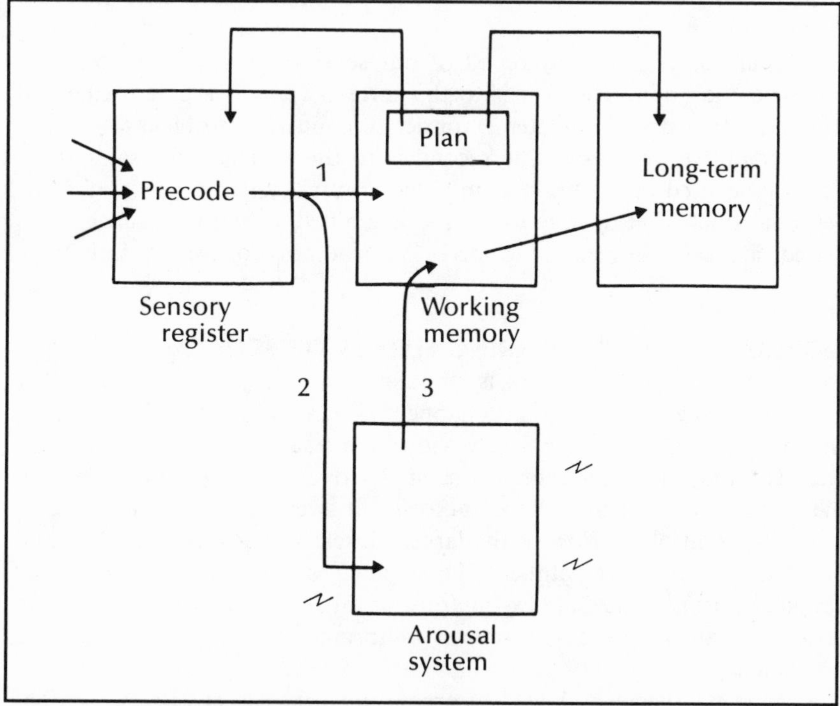

FIG. 4.1. **The Relation between Arousal and Information Processing Systems**

THE ORIENTING RESPONSE

Some messages are more likely than others to be precoded as potentially important. First, any variation from the current input will be precoded as important, for example, a sudden noise against a quiet background or a sudden quiet against a prolonged noise as when the engine stops.

Such variations signal that something different is happening. We are biologically wired up to be sensitive to things that are different, because such variations in the environment may well be vital for survival. Other important signals may include stimuli relevant to a prevailing physiological or psychological set, such as the sound of our own name or the smell of food when we are hungry. Precoding gives a most obvious initial boost to the arousal system when, for example, we hear such phrases as, "Pay attention!," "Fire!," and so on.

There is a special kind of response to unusual precodings of this nature: it is called the *orienting response* because the individual becomes oriented to a new and important event. At the same time, arousal is increased to help him or her handle the new situation. The general picture is given in Fig. 4.2.

Input comes from any or all of our sense organs, and it takes two routes to the brain. The first is to the cortex, where it is interpreted and stored (as discussed in Chapter 2) for decision making, problem solving, or other cognitive processes. The second is to the reticular arousal system, which is located in the brain stem. This arousal system consists of dense neuronal fibers that operate not in a fine-grained, cognitive manner, as do the cortical cells, but simply to increase a readiness for general activity.

ENERGIZING AND INTERFERING EFFECTS OF AROUSAL

When the arousal system is stimulated, as happens in the orienting response, two general things also happen: (1) cortical processes are directly energized to cope more efficiently with the message that has already been received; and (2) the autonomic system is activated to release adrenalin to the bloodstream. There follows increases in sweating, heart rate, rate of breathing, and blood flow to the larger muscle systems; pupils dilate and digestive processes are withheld. These changes are useful for emergency action requiring immediate expenditure of lots of energy, such as fighting or running away. The system returns to normal when the energy has been expended.

However, the general way the organism works can give rise to a problem. Most of the stress we endure is psychological rather than physical and the kind of emergency action required to be taken does not usually demand an immediate, high energy output. Hence we are forced to deal with emo-

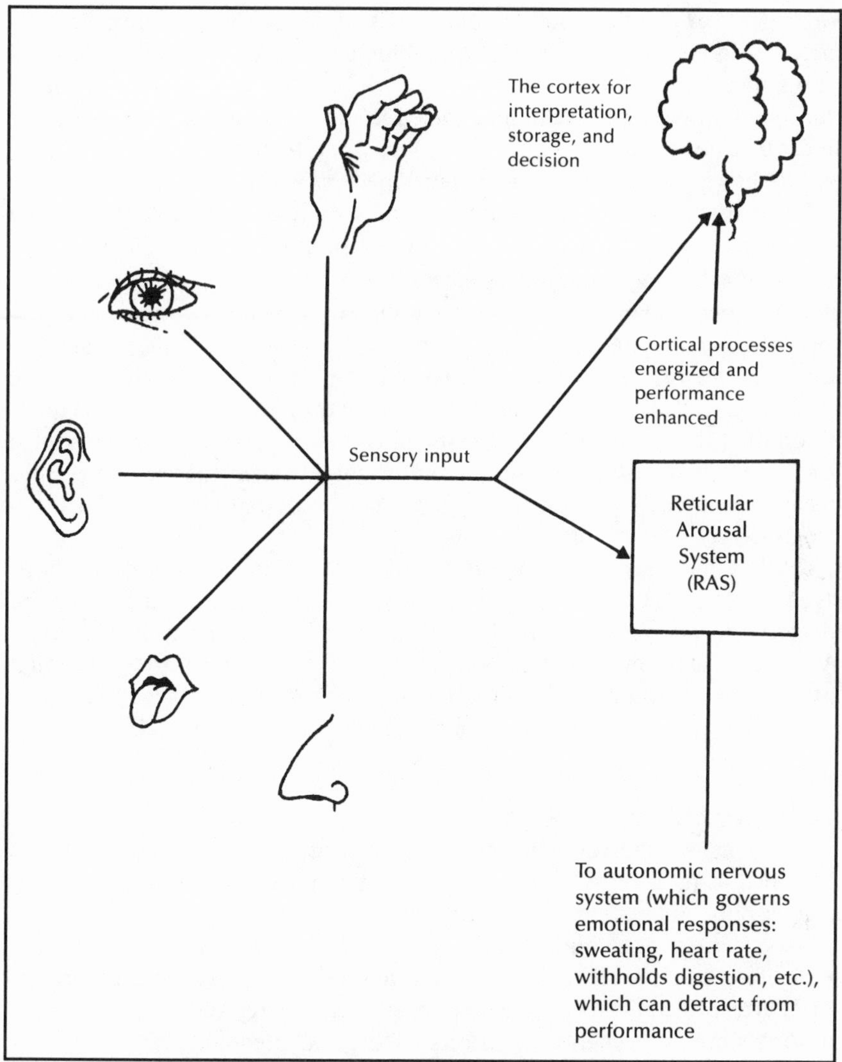

FIG. 4.2. **Relations between Information from Senses, Cortical Processes, and Arousal**

tional or stressful situations with bodily functions that have to be ignored or suppressed in our civilized settings. Moreover, if adrenalin stimulates high heart rate and shuts down the digestive processes, constant stress may lead to either or both of two general kinds of physical breakdown: heart attack and stomach ulcers. People who by nature have particularly unstable or active automatic nervous systems and who live in a stressful work environment, as can be found in some types of flying (or flight instruction), are most likely to be at risk. Continual high stress can cause breakdown in normal people, as has been found in air traffic controllers. But even before that breakdown stage, the immediate effects of such autonomic arousal are likely to distract and interfere with effective performance.

A list of stress factors was developed by T. H. Holmes and R. H. Rahe. It lists sources of stress and assigns a value to each. The total of the values indicates how much stress is present in a person's life (Table 4.1).

Arousal, then, tends to have two effects: an energizing effect, which enhances performances, and an interfering effect, which detracts from performance. The interfering effect greatly increases at higher levels of arousal. One way of illustrating this is to compare the arousal system to the brightness control of a television set, and performance to the picture quality on the tube. If the brightness control is turned down (that is, arousal is low), there is no clear picture. As the brightness is increased, the picture becomes clearer, is clearest at some optimal midpoint, but then becomes progressively washed out.

What happens physiologically is that at optimal levels of arousal, strong signals are simplified and weak signals are suppressed thus giving ideal conditions for creating the clearest and most articulated "picture." When arousal is low, however, both strong and weak signals are given equal attention. For example, trivial matters may be exaggerated beyond all proportion as we wrestle with them in that state just prior to sleep at night.

AROUSAL AND PERFORMANCE

The general relationship between arousal and performance can be described as an inverted U curve. At a very low level of arousal, performance is poor (at extreme levels, of course, we are asleep). As arousal increases, performance improves up to an optimal point (which, as we shall see below, varies with individuals, the nature of the task, and other factors), after which performance deteriorates with increasing arousal. On the upward slope, the energizing effects of arousal predominate; on the descending slope, the interfering effects take over. Anxiety is a common form of arousal.

It is obviously important for instructors to try to strike the best balance, so that students are maximally energized, but the interference point is avoided. A very common example will illustrate: an instructor is questioning a student in a briefing. Gentle probing is likely to provide the right sort of pressure to lead to a good response. However, if the instructor's questions become heavy or sarcastic, the student is likely to become flustered, and if answers are forthcoming at all, they will probably be confused or incorrect.

This example also illustrates some other factors that are involved, such

TABLE 4.1. Stress—Its Origins and Effect

LIFE EVENTS

Rank		Mean Value
1.	Death of spouse	100
2.	Divorce	73
3.	Marital separation	65
4.	Jail term	63
5.	Death of a close family member	63
6.	Personal injury or illness	53
7.	Marriage	50
8.	Fired at work	47
9.	Marital reconciliation	45
10.	Retirement	45
11.	Changes in family member's health	44
12.	Pregnancy	40
13.	Sex difficulties	39
14.	Gain of new family member	39
15.	Business readjustment	39
16.	Change in financial state	38
17.	Death of close friend	37
18.	Change to different line of work	36
19.	Change in number of arguments with spouse	35
20.	Mortgage over $10,000	31
21.	Foreclosure of mortgage or loan	30
22.	Change in work responsibilities	29
23.	Son or daughter leaving home	29
24.	Trouble with in-laws	29
25.	Outstanding personal achievement	28
26.	Wife begins or stops work	26
27.	Begin or end school	26
28.	Change in living conditions	25
29.	Revision of personal habits	24
30.	Trouble with boss	23
31.	Change in work hours, conditions	20
32.	Change in residence	20
33.	Change in schools	20
34.	Change in recreation	19
35.	Change in church activities	19
36.	Change in social activities	18
37.	Mortgage or loan under $10,000	17
38.	Change in sleeping habits	16
39.	Change in number of family get-togethers	15
40.	Change in eating habits	15
41.	Vacation	13
42.	Christmas	12
	LIFE STYLE	
1.	Marital separation	65
2.	Change in responsibility at work	29
3.	Change in living conditions	25
4.	Revision of personal habits	24
5.	Change in working hours or conditions	20
6.	Change in residence	20
7.	Change in recreation	19
8.	Change in social activities	18
9.	Change in sleeping habits	16
10.	Change in eating habits	13

Source: T. H. Holmes and R. H. Rahe, "The Social Adjustment Rating," *Journal of Psychosomatic Research* 11(1967):213–18, in T. S. Holmes and T. H. Holmes, "Short-Term Intrusion into the Life-Style Routine," *Journal of Psychosomatic Research* 14(1970):121–32.

as the nature of the task and the personality of the student. Very simple questions, to which the student knows the answers, are less likely to be affected adversely than complex questions (the answers to which require the integration of a lot of information). Thus a complicated exercise is likely to be failed if the pilot is anxious at the time, whereas a routine exercise would be performed adequately whether the pilot is anxious or not. Low anxiety is a state in which one "couldn't care less," consequently one doesn't bother to try hard. High anxiety disrupts performance; one tries too hard, becomes confused, and the mind "goes blank." With a middling degree of anxiety, however, performance is best for that person, for that task. Complex tasks are best performed at lower levels of anxiety, while simpler ones can withstand quite a high level.

Furthermore, different people are characteristically anxious or nonanxious as part of their basic personality, and consequently react to stress, and cope with stress, in quite different ways.

We have looked at the very general relationship between arousal and performance. Let us now turn to some of the factors that modify the picture.

AROUSAL, TASK COMPLEXITY, AND PERFORMANCE

The Yerkes-Dodson Law, formulated in the early part of this century, states that simple tasks are performed better under higher degrees of motivation or arousal, while complex tasks are performed better under relatively low degrees of motivation. In other words, as indicated in the earlier example of the student being questioned in a preflight brief, a simple question is more likely to be answered satisfactorily under stressful questioning than a complex question. Or to put it another way around, a complex question is not likely to be answered satisfactorily when the student is put under heavy stress. The inverted U curve looks rather different for the two kinds of tasks (see Fig. 4.3).

Line A in Fig. 4.3 graphs a complex task, such as a cross-wind landing. A peak performance is reached quickly with increasing arousal; it lasts a short time, then declines rapidly. Line B refers to a simple task, such as preflight checks, under conditions of increasing arousal; the peak in performance is reached somewhat later, is maintained longer, and then declines relatively slowly. In short, when doing a simple task, a person can operate efficiently at much higher levels of arousal than when doing a complex task.

What do we mean by the terms "simple" and "complex"? Simplicity and complexity are relative terms. What is simple and what is complex directly reflects the amount of working memory that the person needs in order to perform the task satisfactorily. Simple tasks require little working memory, complex tasks rather more. It would therefore follow that any

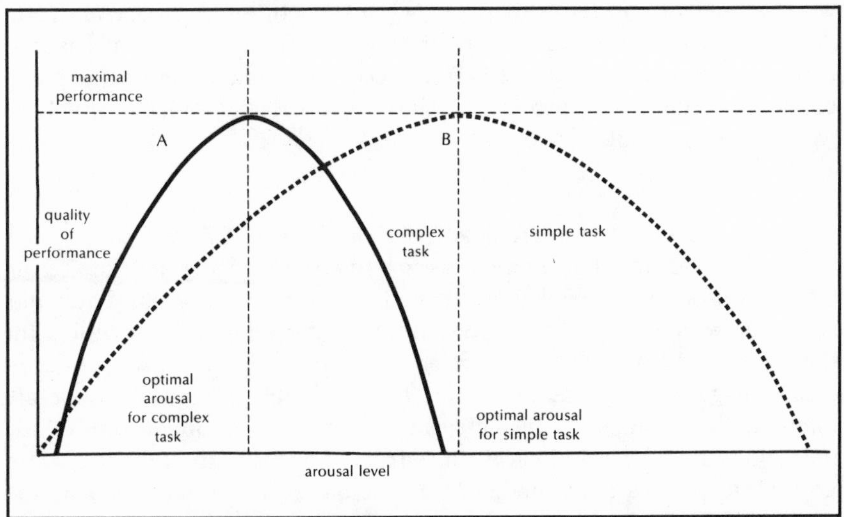

FIG. 4.3. **Arousal and Task Complexity**

ways in which the memory load in performing a task can be lessened would help alleviate the interfering effects of anxiety, particularly in the case of complex tasks. Mnemonic devices, such as those we looked at in Chapter 2—diagrams, notes, outlines, and other ways of organizing complex ideas into smaller, manageable chunks—would be useful for counteracting the anxiety experienced in performing a task. Conference speakers, particularly those who are at all nervous, find notes very useful for countering anxiety, even if they never use them in the event. An anxious flight instructor may find the use of an overhead projector could assist to allay fears of a mass brief.

Apart from these external props, however, there are two processes that are vital in freeing more space in working memory: coding and rehearsal. Rehearsal and coding play complementary roles in the learning of complex material. Rehearsal is used predominantly in the learning of skills, coding in the learning of meaningful material. However, even in the latter case, rehearsal is necessary to "fix" the learned codes.

Let us take the more direct case of skills first. Experienced pilots have automated their reactions, primarily by rehearsing them frequently, so that they do not need consciously to think about them except in an emergency. The working memory of beginners, on the other hand, is fully occupied with the sheer mechanics of control, so that there is no room available to enable them to attend to further refinements. Likewise, in the case of coding, we saw that generic codes greatly decrease the quantity of drain on working memory. In terms of learning the skills of aircraft control, a student needs to reach the autonomous stage (described in the last chapter) before working memory space is freed. Until that time, the working memory will be engaged in providing solutions to the manipulative problems with aircraft controls.

HOW AROUSAL INTERFERES WITH PERFORMANCE

We should now link this discussion to the energizing and interfering effects of arousal. Remember that the energizing effect of arousal accounts for the upward slope of the curves, and the interfering effect accounts for the downward slope of the curves as depicted in Fig. 4.3. As arousal increases it generates more autonomic or emotional stimuli, and these are internal. For example, with increasing anxiety, we become aware of the physiological and psychological side effects of arousal. In other words, we notice that our heart is pounding, we have butterflies in our stomach, and we feel helpless. These general I-feel-awful cues take over working memory to the exclusion of task-relevant information. However, because complex tasks require more task-relevant cues for adequate processing than do simple tasks, task-relevant cues will be displaced earlier when performing a complex task than will be the case with a simple task. The performer in a complex task, such as landing the aircraft, will in effect be operating with reduced working memory, and cannot afford to do that in a complex task. Hence performance suffers. In simple tasks, on the other hand, reduction of the space will not matter so much, and so energizing (improvement) will continue for longer. Fig. 4.4 illustrates stress level in relation to the various phases of a flight.

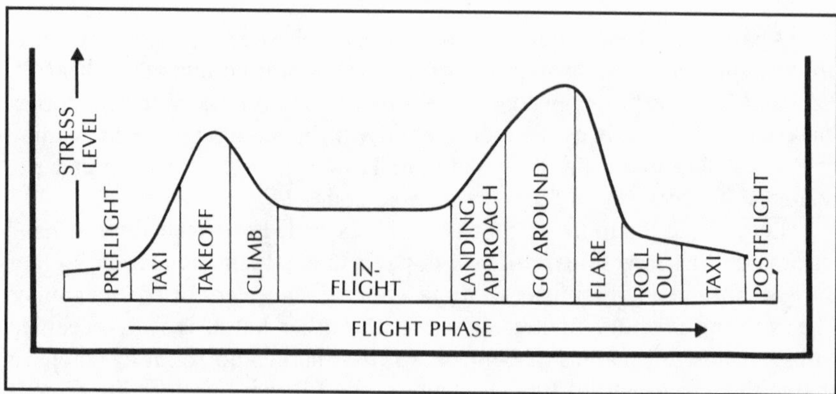

FIG. 4.4. **Stress from Pilot Workload**

PERSONALITY DIFFERENCES IN AROUSAL

We have noted that people differ markedly in their reactivity to stress: some people overreact, and hence tend to seek quietness and solitude in order to work comfortably, while others take stress in stride, and indeed seem to go out of their way to seek excitement and stimulation.

Psychologists usually make a distinction between trait anxiety and state anxiety. *Trait anxiety* refers to a general readiness to react with anxiety to most situations. *State anxiety* refers to the anxiety actually experienced in a given situation.

A person who is high in trait anxiety is described in everyday terms as "nervous," "jumpy," "uptight": it is part of the way that person is. However, with experience he or she learns to focus feelings of anxiety, and often the target of anxiety changes. If he or she has a history of success, then a particular situation that once caused enormous anxiety may become quite innocuous, frequently enjoyable.

State anxiety is specific to a situation. The person feels strong anxiety only when in that situation. A form of state anxiety is "test anxiety," which increases when a student experiences repeated failures in tests and examinations. Test anxiety is almost always debilitating, preventing students from performing to the best of their ability. Student pilots, for example, may perform perfectly: until the instructor tells them that they are going on a test flight.

Trait anxiety, or general anxiety as it is sometimes called, is a characteristic predisposition of a person to react with anxiety. It is very likely part of the person's physiology, particularly of the reticular arousal system and of the autonomic system, and is probably inherited—which is not to say, of course, that people do not learn their own ways of keeping their general anxiety level under reasonable control. Sometimes these ways are useful and help them to adapt, for example, finding out more information about a potential threat. Other ways, however, are maladaptive. The person may learn to cope with a feared test, simply by avoiding tests, being "ill" when the examinations are on, dropping out of flying school, or allowing a student license to lapse.

Research results indicate that trait anxiety affects performance only to the extent to which it is actually experienced. People low on trait anxiety will need a large "push" before the anxiety induced by the environment will be aroused to the extent that it will affect them adversely. People easily aroused to anxiety, on the other hand, only need, as these experiments show, to be told that the result is "important," or to find that the items being tackled are "difficult," and the anxiety thus aroused will inhibit performance.

When anxiety is associated with a particular set of circumstances, we speak of state anxiety. For example, test anxiety is associated with anxiety about doing poorly when being evaluated. The student's first encounters with flight tests will inevitably be associated with some test anxiety.

In flight instruction, teaching takes place from what one experienced pilot described as "the noisiest, most cramped, uncomfortable learner's seat in the world." He argued that neither books nor a commercial license make a good instructor. The key, in his view, was motivation to teach, and to be good at it. Being good at it involves the recognition and alleviation of stress in the student and oneself.

For example, an experienced instructor has a personal cue for the onset of stress: he finds himself starting to whistle. His solution is to deliberately change his posture by shifting in the seat and establishing a new body position. As a possible solution to student stress, he advocates the conventional break from instruction with the instructor taking over the controls, plus a more novel approach. This involves asking the student to use finger tip and thumb contact on the control column. He sees it as a cure for the knuckle-whiteners.

Another instructor suggests the use of a five-minute "free" period when stress begins to interfere with student performance. During this time the student is free to fly the aircraft as desired. Apart from the opportunity to relax, the period often provides insights into student progress for the observant instructor.

WHAT CAN STUDENTS DO ABOUT STRESS?

Several instructors have provided suggestions for helping students handle stress associated with flying. Most common is helping the student pilots to talk about matters of concern and to clarify aspects of the instructional situation that are really out of their control. For example, the weather could be quite demanding or traffic may build up unexpectedly, providing a situation too complex for the learner's current stage of development. Students can be encouraged to take time out for relaxation: fishing, jogging, sports—simply doing something they really enjoy. Physical health and psychological well-being can be boosted simultaneously. One instructor described a case in which a student pilot was using medication to counter symptoms of anxiety, and makes a point of advising students against this practice.

Finally, a number of instructors pointed out the link between a positive self-concept and low anxiety. If students are asked to list their attainments as opposed to their failures, the balance will usually favor the positive. When they feel better about themselves as pilots, a complex situation becomes a challenge, and is not perceived as stressful and the likely source of failure.

Test Anxiety

People who are in a state of anxiety on a test respond to their own feelings of anxiety, and these responses interfere with the response they should be making to the test.

A particular response to test anxiety is fear of failure. Test-anxious individuals are preoccupied by the possibility that they will fail, and think that that would be a catastrophically awful thing to happen to them. Consequently, attempts at coping are directed towards avoiding the catastrophe of failing. But they are in a bind. Since past history has shown that success in the task is unlikely as they have failed in the past (which is precisely why they are test anxious now), they will adopt strategies that help them avoid failure but that do not necessarily, or even probably, involve success. This leaves them with a limited range of options, none of which can be regarded as very helpful, and that might include: migraine (or some other, particularly psychosomatic, illness) on the day of the test; tackling only the very hard or the extremely easy items (but none of the ones they might reasonably be expected to handle as a good test of their abilities); or finally, just dropping out, which, by this stage, might well be the best decision they can make. However, there are adaptive ways of handling the situation, as outlined below.

COPING WITH TEST ANXIETY

Those who do feel anxiety in tests and exams tend to do worse than those who do not. This is not very surprising. Lack of success is one of the reasons that anxiety is experienced. Usually the anxiety so experienced is sufficiently strong to go beyond the optimal level to the point where it interferes with performance.

State anxiety creates three classes of response: (1) bodily reactions such as sweating, palpitation, discomfort; (2) thoughts about one's inability to cope with threat; and (3) coping styles that have been effective in the past.

All these demand space in working memory: one becomes aware of physical discomfort, if it is severe enough and also of such questions as "What am I going to do? What can I do?" Unfortunately, such thoughts are irrelevant information in the sense that they will not lead to a solution of the problem. However, they do not go away. They crowd memory, and force out relevant information. In general, and particularly where heavy demands are made on working memory (which are increased under timed or speeded conditions), anxious students do worse than nonanxious.

However, anxiety also has an activating or arousal function as well, which leads to rapid processing. If the interfering effects can be prevented, then one is left with the beneficial effects of activation. The clues are therefore (1) to make the items of relevant information as stable as possible, so that they will resist the demands of irrelevant information for working memory space; and (2) to reduce the load on working memory as much as possible.

Anxious learners should train themselves to use notes, maps, diagrams, outlines, and the like, wherever possible. For example, in a theory examination if students have a thought that is relevant to another part of the same question, or even to another question, they should note it down immediately. To try to remember an isolated point will result either in forgetting that point or in losing track of one's present train of thought. When they first read the test question, candidates are advised to jot down whatever points occur to them, and then to make up an outline, in logical order, with subheadings. They may then turn back to answering the question itself. If they have learned the material well, once they start, relevant information will come rushing back by association. This relevant information will in turn crowd out such irrelevant thoughts as: "Gosh, how am I going to do this? I feel quite sick. . . ."

If this process is to work, however, there must be a solid background of well-learned material. It is unlikely that such a background can be acquired in the week or two just prior to the exam. If one's notes and outlines are based on just a week of rote learning, then the immediate associations that have been incorporated and written down will be quickly exhausted with nothing to replace them.

Generic coding is the answer. A generic code is like a fishing line with many short lines and hooks attached and to each hook is attached another line, also with many hooks. By hauling in the main line, a great deal of richly associated material is drawn into working memory. If it is sufficiently well learned, such material will displace the feelings directly due to anxiety itself.

The general strategy of such learning is cumulative and occurs in four stages:

1. Learning. The initial learning itself is important: a book chapter, a lecture, or whatever. The basic material is read and assimilated. Consider, for example, a student's preparation for attitude instrument flying. Reading and teaching will cover the physiological aspects, the instruments, aircraft control by cross-reference, pitch control, direction and lateral control, the flight maneuvers, recovery from unusual attitudes, spinning and recovery, instrument flight after engine failure, and changing from instrument flight to visual flight and vice versa.

2. Elaboration. The learner thinks about what he or she has learned, ruminates, and relates it to what other relevant material the learner already knows. The learner will probably learn some new material, which helps reevaluate the old.

For example, the student will be relating airspeed indicator errors (instrument, pressure or position, compressibility, density error, blockages or leaks) to airplane systems study completed in ground school or theory lessons. Similarly, the student will link the effects of inertia in attitude instrument flight with prior knowledge and experience of the forces involved in flight.

3. Organization. Gradually a pattern emerges. The incident, or piece of content, "makes sense." To take an example from this text, the concept of working memory at this stage emerges as more than a flashlight pattern on a dark night, but part of a general process of learning and memorizing, which is also involved in countering the effects of anxiety, and so on.

At this stage the student may begin to visualize the desired flight performance, and this "vision" will enable appropriate selection of attitude and power. As instructors appreciate, this process takes time. With theory, the process is hastened by much note taking during both elaboration and the next stage, consolidation.

4. Consolidation. Learners start making notes of their theory notes, interrelating all the bits and pieces about working memory, for example, that they have picked up in the references, in discussion, and in their original classwork. They then make notes of these notes, use spatial summaries, and so on. Finally an outline of the whole course may almost literally be written on a bus ticket. At that point, learners are ready to rote learn the notes-of-notes—not to understand anything any better, but simply to make sure they can recall those notes on cue. However, because of the background of learning, a particular word or pattern in the notes is like one of these hooks on the main line: it is attached by cross-referencing to so much other relevant material.

At this stage in instrument flight the student is automatically using the control technique of change-check-hold-adjust-trim and has sufficient relaxation to maintain a light but positive control.

People who study using this four-stage process—learning, elaboration, organization, and consolidation—can even afford to be in a state of high anxiety during a theory exam. Once they get going (and that is the important thing), that richly coded information will almost literally take over. The student then has the advantage of working under conditions of high arousal with a working memory dominated with relevant information. Indeed, this is the recipe for high intrinsic motivation and high quality work.

Arousal and Variability in Instruction

As we saw in Chapter 2, attention can be aroused and maintained by the use of factors that are intrinsic to the material or by using organizers, overviews, and questions that relate meaningfully to the material. Unfortunately, it is not always possible to use them. For example, students may not be interested in anything of potential educational value, as far as a particular lesson is concerned.

An alternative way of handling the situation is to make such attention grabbers extrinsic to the material. Television commercials use the technique, as do directors of feature films, and even flying instructors. In a mass brief on principles of flight, the instructor walked into the room, waited for silence, and beckoned to a student. The student at the front of the room was given a dart and asked to throw it at the board. Lacking flights to stabilize its trajectory, the dart thudded into the board side-on and dropped to the floor. Why? The lesson had begun.

This incident clearly illustrates an element in creating high arousal: surprise. Surprise is the result of an unexpected happening. Very simple physical examples include variation in the intensity of a stimulus. Like surprise, a variation can be defined as a departure from the expected. The skill used by teachers who deliberately introduce variations in lessons is known as *variability.*

Although our expectations can be defined in terms of any kind of dimension or attribute that may be operative, variability can be considered in terms of three components: the instructor's manner or style, the media and materials of instruction, and the interaction between instructor and student.

STYLE OF INSTRUCTION

The variations in an instructor's personal style are essentially either verbal or nonverbal in nature.

Verbal Variations

To be "interesting" speakers need to do such things as vary the speed of delivery, slowing down for emphasis; increasing volume for importance; or, a double surprise, speaking the climax very softly so that the audience strains to hear. The instructor's voice can range from sharp emphasis to quiet encouragement. Tone and pitch provide further contrasts with volume and speed. Variation of grammatical style and use of questions at unpredictable moments during the lesson are further examples of formal or extrinsic ways of creating the unexpected.

Nonverbal Variations

Nonverbal variability can reinforce verbal techniques. As a student struggles to identify the error in the solution of a mathematical problem, the instructor silently goes to the board and underlines, in colored chalk, the key step.

Nonverbal communication can take the form of facial expressions of approval or disapproval, head or body movements conveying the same message, and gestures. The movement of a teacher around the classroom is itself a form of variation and nonverbal communication. The manipulation of eye contact with students is another form. Another aspect of variability bridges the verbal and nonverbal. This is the versatile technique of pausing for such purposes as capturing attention, segmenting the lesson, or giving students time to ask questions or organize comments and their answers. It is also a vital part of teaching in an aircraft cockpit where a student will quickly seize upon any nonverbal communication (such as anxiety) from the instructor.

MEDIA AND MATERIALS OF INSTRUCTION

Basically, there are three types of teaching material: visual, aural, and tactile. All of these have variations.

Apart from actual objects, pictures, maps, charts, and chalkboard illustrations, visual resources include film (8mm, 16mm, and 35mm), videotape, television, overhead projector transparencies, and duplicated materials. Some or most of these are available to flight instructors, but there is also the opportunity to use the reality of the flying school's environment. Field trips and excursions to the airport, aircraft, control towers, and so on provide a valuable opportunity and add purpose, variety, and interest to preceding and following lessons. How much preferable it is to see the aircraft stripped for a major overhaul, than to examine diagrams or individual parts in the classroom.

In varying proportion, teacher talk and pupil talk are the main aural components in a lesson. Teacher talk, according to research findings, comprises one-half to two-thirds of all classroom interaction time. Given that situation, perhaps instructor talk is the first component to consider for variation, especially if student interest and surprise is the criterion. In flight instruction, however, there can also be a place for recordings of sound or interviews.

It is fairly common practice in ground schools for actual parts of an aircraft to be handled. Cross-sections, scale models, and topographic forms can all be used for tactile variations.

INTERACTION OF INSTRUCTOR AND STUDENT

The final source of variation has more limited appreciation in flight instruction in which most teaching is in the one-to-one situation. This interaction, however, can be considered as a continuum.

At one extreme is instructor talk, at the other is independent work by students. In between are such variations as group work, student presentations, flight instructor assistance of individuals or groups, and so on. Interaction offers the instructor a means of maintaining surprise. The mass brief begins with the instructor introducing the subject; then groups are formed to discuss it and reach a conclusion, which individuals convey to the whole group. Such a lesson has a smooth and purposeful change of interaction every 10 or 15 minutes.

These three components of variability share a common purpose: departures from the expected to provide learner surprise. The student's surprise is the link with motivation to learn. Surprise provides both the signal for precoding the appropriate cue and the arousal boost needed for continuing attention. However, the instructor should not go overboard. We have seen lessons taught with simultaneous use of a battery of high-powered stereo speakers, a tape-slide presentation, three or four overhead projectors, a couple of color VTRs, a strobe light or two, and some recorded sound effects. One can guarantee surprise with that devastating array, but the aim after all, is student learning, not mind blowing.

QANTAS FLIGHT OPERATIONS TRAINING

QANTAS flight operations training provides some insights into the use of variability. Operations training covers ground training responsibilities in four areas: (1) pilot and flight engineer training on aircraft systems, their control, and operation; (2) flight crew safety training, which covers pilots, flight engineers, and flight attendant training for aircraft evacuation procedures, ditching procedures, and inflight emergencies; (3) licensed aircraft maintenance engineer training; and (4) apprentice training program providing on-the-job experience in servicing overhaul.

Work experience record books enable the apprentices to be moved through various areas during the four years of apprenticeship to ensure varied and wide experience. Personal development is considered by incorporating nonengineering work experience in sections, such as marketing, finance, personnel, communications, computer services, and flight control. Apprentices undergo an Outward Bound course as part of their training, and also enroll in units on supervision, communication skills, and decision making.

In QANTAS, a training research and development group is formed from experienced instructors drawn from each training group. This unit has the responsibility of research and development for new training methods, techniques, and training equipment. It develops training tape/slide kits, video films, and hands-on training procedures. A production department consists of a photographer, color processing equipment, sound studio with mixers, dubbing equipment and high-speed tape copiers, and word processor (for scripts, teaching outlines, and so on). One hour of classroom presentation, using dual 35mm projectors synchronized by a cassette tape player presenting a scripted lesson, involves between 250 and 500 hours in development, production, and review. Additionally, tape scripts will frequently be reproduced in text form with multiple choice questions. These study guides enable a review of the day's work and incorporate the visuals previously conveyed by the 35mm slides. The development team is currently working software for computer-aided and computer-managed instruction.

QANTAS has nine specially designed classrooms, each having three projection screens (two for 35mm slides and one for 16mm projection. The latter doubles as a writing board.). The writing screen is limited to background information, such as the course number, instructor's name, and so on. All other information should be provided in the training package. The instructor controls all projectors, lights, and a student-response monitoring system (in which students select alternatives to questions provided after approximately each 20 slides). The QANTAS philosophy is to train people in their working environment (such as the aircraft cabin) and restrict the classroom to the provision of information about course objectives, overviews of tasks and skills required, and aircraft systems information.

Other QANTAS training facilities include tutorial rooms, carrel rooms, evacuation trainer, ditching trainer, inertial navigation system trainer, B747 ground simulator, and full-flight simulators.

Summary and Conclusion

Behavior does not take place at all unless it is motivated. Motivation has two aspects: (1) general drive or energy level governed by the reticular arousal system; and (2) specific aspects that refer to the individual performing this particular task and no other. This chapter looks at the general aspects; the next looks at specific aspects.

Stimuli in general, and particularly varying stimuli, act both upon the cortex and the arousal system. Increasing arousal has two effects: it energizes cortical functions; and it causes, increasingly at high levels, an autonomic or emotional response that may interfere with cognitive functioning.

Interference effects are affected by difference between tasks and differences between individuals. Task differences are determined by the amount of memory load the task makes on the individual, which in turn is a function of the individual's familiarity and skill with the task. Simple tasks are performed effectively at much higher levels of arousal than complex tasks. Minimizing the impairment of complex tasks is to a large extent a matter of automatizing the individual's responses as far as possible, thus decreasing memory load.

Individuals differ widely in the way in which their arousal systems react to stimulation and in how they themselves handle changing in arousal. Trait anxiety refers to readiness to react with anxiety. State anxiety refers to experienced anxiety in particular kinds of situations.

The general drive to behave involves the arousal system, which affects behavior in two different ways: orienting at the beginning of the task, and facilitating or interfering throughout the task. The orienting response is essential in teaching technique and three aspects of variability were discussed: verbal and nonverbal styles of teaching; media and materials of instruction, capitalizing on variations in sensory input; and modes of pupil-teacher interaction. Such variability is extrinsic to the content being taught. It has nothing to do directly with the message itself. Nevertheless, surprising media and contexts can be very powerful in attracting the learner's attention at the right time and focusing it onto the right track.

The effects of arousal on performance for the duration of the task depend on: the personality of the learner, the stressors in the immediate situation, the nature of the task, and the learner's previous history of success on that task. People have, probably genetically, a propensity to react with more or less anxiety: they are called trait anxious or neurotic. Anxiety only becomes a functional issue if it is felt, as a particular state, in given situations. Such state anxieties include test anxiety and teaching anxiety. Both are likely to be encountered in flight instruction.

Discussion Questions

1. Using the stress table (Table 4.1), calculate the extent of the stress one could typically expect to find in (a) an *ab initio* student pilot, (b) a commercial pilot, and (c) a junior flight instructor. Justify your analysis.
2. Briefly indicate what you consider to be the implications of these ratings.
3. (a) When flying, what are the symptoms that enable you to identify stress in a student and in yourself? (b) What remedies do you apply to remediate the situation?
4. You are planning a preflight briefing on the primary effects of controls. Show how variability can be introduced through the media of instruction by utilizing the following: models, overhead transparencies, 35mm color slides, video or movie, student observations, duplicated sheets, personal reading, sketches or diagrams, computer assisted instruction, programmed instruction, magnetic board, chalkboard or whiteboard, photographs.
5. In a mass brief on the same topic, how can interaction be used as a source of variability?

Further Reading

Biggs, J. B., and R. A. Telfer. *The Process of Learning*. Sydney: Prentice-Hall, 1987. Chapter 3 is devoted to motivation and academic learning. Suggested additional reading includes three titles dealing with drives and the central nervous system, stress and achievement, and anxiety and educational achievement.

Hockey, R., ed. *Stress and Fatigue in Human Performance.* Chichester: Wiley, 1983. Chapters by specialist contributors from Europe and the United States cover the effects of monotony and boredom; heat and cold; noise; circadian rhythms; fear, fatigue, incentives, stress, and drugs; anxiety and individual differences; and coping efficiency and situational demands. Hockey draws attention to the need for more information on coping strategies and stress effects.

Turney, C., et al. *Sydney Micro Skills: Series 1 – Reinforcement; Basic Questioning; Variability.* Sydney: University of Sydney Press, 1983. Part of a series of teaching skills handbooks, this one is accompanied by a recently redeveloped videotape. Oriented to school teaching, it retains application to vocational instruction. An interesting project for a CFI would be to develop a similar video suitable for use in instructor refresher courses.

5 Motivation and Self-concept of Student Pilots

THE FOLLOWING QUESTIONS are answered in this chapter: What is the Premack principle, and how can it be applied in flight instruction? Is punishment the same as discipline? When and how often should a student be praised? How much influence does a flight instructor have on a student pilot? What's most important for students: the need to achieve success or the need to avoid failure? How can conflict in ideas motivate learning? How does attribution theory relate to a student's success or failure?

After reading this chapter, you should be able to (1) clarify the motives of your student pilots, (2) distinguish between extrinsic and intrinsic motivation, (3) structure flight instruction exercises to condition overanxious or timid students so that they can enjoy flight, (4) apply the Premack principle to flight instruction, (5) assist a student to want to learn, and (6) recognize the role of a student's self-perception regarding efficiency in determining flight performance.

Motivation

Why do people want to learn to fly aircraft? For the same sorts of reasons that they do most things in life: because it is a necessary step in achieving something else they want; and/or because flying is enjoyable.

As we will see later in this chapter, these reasons are not mutually exclusive. The first refers to extrinsic motivation, that is, flying is a means of achieving a payoff, such as material gain, social approval, or ego enhancement. The second reason refers to intrinsic motivation; the sole payoff is the sheer enjoyment of flying.

If you refer to Fig. 5.1 you will see that along the baseline there can be a range of motivations between the extremes of "pure" extrinsic and "pure" intrinsic. We will now examine these and relate them to the ways a flight instructor can interact with the student. First, however, let's establish some basic concepts in this area of motivation.

Extrinsic Motivation

CONDITIONING THE EMOTIONS

If you have read Anthony Burgess's novel *The Clockwork Orange* or have seen the movie, you will remember that Alex, the young antihero, was committed to violence, sex, and classical music. In an experiment in which films of Nazi atrocities were accompanied by a nauseating drug, his behavior changes. As a result of conditioning by association, violence and sex now make him ill. Unfortunately, though, music, too, nauseates him, because Beethoven's Ninth Symphony was background music to the atrocities film—but that's a different part of the action, as far as we are concerned.

This kind of conditioning is called *classical conditioning*. In flight

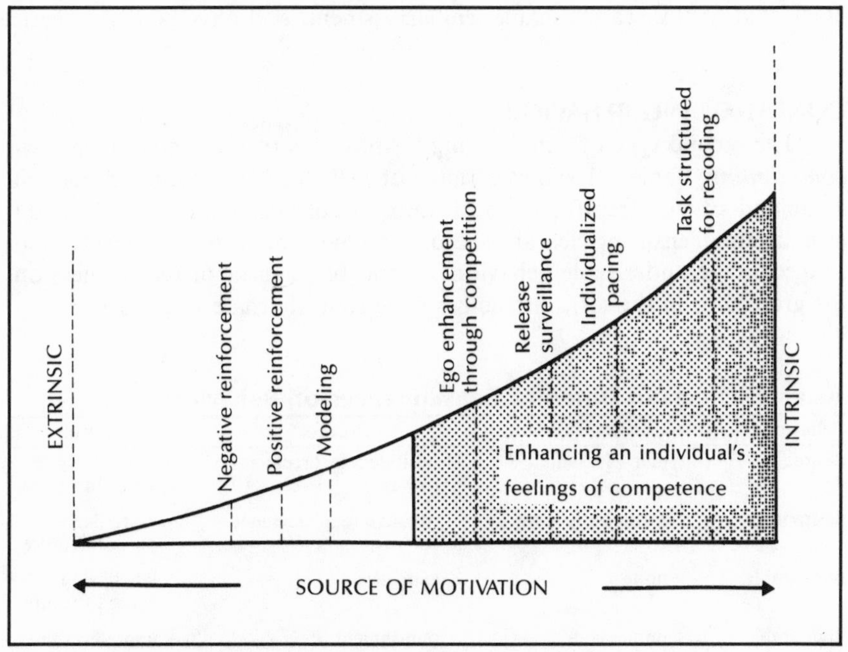

FIG. 5.1. **Sources of Student Motivation and Self-concept**

instruction, it can be seen in the plight of the overanxious and timid student who needs to be given the opportunity of associating other emotions with the experience of flying, such as savoring the company of a soaring eagle (at a safe distance!), or capturing the imagery of the mist on morning mountains. The instructor needs to give time off from steep turns, stalls, or spins and provide an experience that will have pleasurable associations. ("How did the lesson go today?" "Great, you should have seen the mountains this morning. We were up at 4000 feet and. . . .") Otherwise, the stimulus (flying sequences) becomes associated with an inappropriate feeling (lack of confidence, nausea).

Another example can be found when a student has failed to complete prescribed theory assignments. Punishing the student by providing additional assignments or problems in navigation or meteorology is clearly poor psychology. The assignments are associated with punishment and an intensified dislike of the subject results. (Remember Alex's Beethoven?) One of the authors once taught with an English teacher who punished students by making them do math problems. Perhaps flight instructors set penalty assignments in ground school subjects!

Classical conditioning provides an explanation for apparently irrational dislikes of student pilots for theory subjects, individual flight instructors, or specific flying sequences causing discomfort. Instructors can be associated with sarcasm, shame, embarrassment, and physical discomfort.

CONDITIONING BEHAVIOR

The second type of conditioning is known as *instrumental* or *operant conditioning.* Table 5.1 shows a range of influences on a student pilot in command of an aircraft. To inculcate desirable behavior (such as good airmanship in ensuring clearance before commencing turns or descents) and to discourage undesirable behavior (such as blind turns or recklessness on the ground or in the air), instructors have four alternate responses.

TABLE 5.1. **Effects of Different Consequences on Behavior**

Behavior	Consequence	Process	Result
desirable	gain pleasant	positive reinforcement (reward)	more likely to occur in future
desirable	avoid unpleasant	negative reinforcement (reward)	more likely to occur in future
undesirable	none	extinction	less likely to occur in future
undesirable	unpleasant	punishment	unpredictable except in special circumstances

Let's say the student consistently fails to remember to change fuel tanks. The instructor can seize the opportunity to praise the student if and when the student does remember (see "positive reinforcement," line 1 of Table 5.1). This makes it more likely that the student will remember to change fuel tanks in the future.

On the other hand (see line 2, Table 5.1), the instructor may not praise the student, but could point out in the preflight brief the importance of all checks and the fact that the student had been forgetting the fuel change-over. While taxiing, the instructor could also comment in some way. Then, when the aircraft is parked ready for run-up, the instructor significantly says nothing. Anxiety becomes the negative reinforcer. If the appropriate response is carried out, the threat is removed. The weakness with this method is that the use of negative reinforcement could associate anxiety with the total context of the air exercise and with flying.

The third alternative (line 3, Table 5.1) is not to increase the response (of switching tanks) by reinforcement, but to extinguish the response (of failing to switch tanks) by providing no reinforcement. When the student remembers, the student is praised. When the student does not (assuming safety is not jeopardized in any way), the matter is not explicitly raised. The point is that instructors may tend to ignore all the right things a student does and comment only on the errors. They expect thoroughness, and naturally find it difficult to ignore what they do not like or expect.

The last alternative (line 4, Table 5.1) is that of punishment. If the student forgets the fuel tank switch, he or she can write out the pretakeoff checks a hundred times, or take an extended preflight brief instead of the next exercise, or can't solo that day, or. . . . You know the sort of thing. Punishment is an unpleasant consequence for failing to switch tanks. The punishment comes from another person—the instructor.

At this stage, let's move from the example of switching tanks to more general implications of these four approaches to behavior modification or conditioning methods.

CONSEQUENCES OF STUDENT PILOT BEHAVIOR

There are, then, four different kinds of consequence to a student pilot's behavior. Each has quite different implications with regard to the likelihood of that behavior occurring in the future.

Positive Reinforcement

Some positive reinforcers, such as consumables (food, drink, sweets, etc.), money, or tokens (which can later be exchanged for a real reinforcer) are more applicable in the home than in flight instruction. But some classes of positive reinforcers are relevant.

Social reinforcers can be verbal or nonverbal. These convey the approval of the instructor for the student's response or action. Phrases like "Way to go!" "Right on!" "You're doing fine," are examples. Nonverbal reinforcers include paying attention, listening closely, nods of assent or agreement, admiring or praising glance, and so on. Such nonverbal reinforcers can be more powerful than verbal reinforcers. In flight instruction, however, social reinforcement depends for its value upon the relationship between instructor and student. If the student neither likes nor respects the instructor, then praise can become unpleasant and patronizing, not rewarding.

The Premack principle, named after psychologist David Premack, suggests that a preferred activity (such as flying an aircraft) provides a very effective means of reinforcing a less-preferred activity (such as learning checks and procedures). If a student does one thing a lot, presumably because of a love for that activity, it is reinforcing to be given the opportunity to do it. Thus, students with theory subject problems could be encouraged to devote more time to their study by utilizing their enjoyment of a particular aspect of flying, such as solo periods, low-level flying, traffic patterns, or upper air exercises. The rule is that there is to be none of that flying until the theory standard has been attained.

Negative Reinforcement

As indicated above with the switch of tank example, negative reinforcement does not mean punishment. Rather, it means the avoidance of punishment. Negative reinforcement is where the consequences of the desired response remove distress. It is thus rewarding rather than punishing.

Consider the part anxiety plays in hindering a student pilot's learning. The anxiety is generated by a threat of punishment. Perhaps most of a student's peers have flown solo, and the student is way above their average number of hours dual. That student's anxiety is generated by the threat of loss of peer esteem. The reward is the alleviation of the anxiety. The student goes solo, and the group gathers at the airport fence. Their jeers of derision about the bounce on landing are music to his ears. He's flown solo! The appropriate response was carried out; the threat is averted.

Extinction

Reinforcement increases response strength; no reinforcement extinguishes the response. Unlike other educational forms, flight instruction probably can't afford the luxury of extinction by ignoring the response as suggested in the omission of a check before takeoff. It's one thing to ignore the four-year-old's swearing: but consistently missed checks or out-of-balance flying are other matters.

Punishment

Discipline has traditionally played an important role in flight instruction, especially in flying schools that seek to convert young men and women into commercial pilots. Punishment can be distinguished from discipline. One psychologist, William Glasser, argued that discipline involved pain as a natural consequence of one's behavior. Punishment involves pain caused by someone else's disapproval. In the first case the onus is on the student ("I knew I'd fail the emergency landing if I didn't practice it in my solo sessions. I blew it!"), but in the second, it's on the punisher ("He's got it in for me. He's never going to pass me."). Obviously the student's future behavior is going to be very different, depending upon whether he or she reads the situation as involving discipline ("my fault") or punishment ("instructor's fault"). With self-discipline as the ultimate goal, onus on the student is the guiding principle.

In Fig. 5.1, sources of motivation are traced from the extrinsic to the intrinsic. In other words, the knowledge of checks can be motivated from anxiety at the consequences of missing a check (negative reinforcement); being able to perform the checks just like the instructor or commercial pilots (modeling); doing the checks better than other students (ego enhancement through competition); showing the ability to do the checks without any need of outside supervision (release from surveillance); doing the checks better than he or she had ever done previously (individualized pacing); and doing the checks perfectly as just one aspect of complete preparation for and accurate execution of the flight exercise.

The progression here is from an external control of motivation by reference to the instructor or others, to the other extreme of the individual student setting up standards and enhancing his or her feelings of competence.

This is an important aspect of learning and especially of flight instruction. As we will see the extent to which students attribute success or failure to themselves or to others is an important factor determining success or failure.

SOME ASPECTS OF REINFORCEMENT

Timing

Reinforcement should follow the response as soon as possible. There's not much chance of ecstatic responses from students when you remind them two weeks later of their excellent cross-wind landings. But after touchdown it would have been different.

It's very easy for training pilots to forget to apply this principle in flight instruction, because the instructor is usually mentally in front of the aircraft (and the student's immediate response), planning the next aspect of the flight.

How Often?

Once the correct response has been established, reinforcement is usually most effective on a partial basis. The maintenance of coordination in turns or with power changes is not praised every time it is implemented, but every so often. Like the reward of solo-time for a job well done in the dual exercise, students never are quite certain when it will occur. Like the slot machine addict, they keep plugging away. They know that the instructor recognizes meritorious performance, because they have evidence of that recognition. When recognition is again given, that's enough evidence to keep them striving for it again the next time.

Self-reinforcement

Here students promise themselves a night on the town if they pass the test, or 15 minutes free flying if they complete their solo practice to required standard. When studying, instead of making coffee before one starts, reinforcement principles indicate that it would be more effective to set a target of 2 hours or 10 problems, and then have coffee.

SOCIAL MOTIVATION

One of the most powerful sources of influence upon a person's behavior is another person. We do things because it is important to us that we appear favorably in the eyes of others. Perhaps because we feel good inside, people tend to imitate or model others even when not directly rewarded for doing so. It boosts the self-concept to do what other important people do. A self-concept is the image or concept people have of themselves, especially of their physical, social, or mental abilities, and the values they place upon these self-evaluations.

Models tend to be the most liked and respected: individuals will model those they would like to resemble and can identify with. Instructors beware! Student attitudes toward danger, alcohol, fast cars, coffee, and tobacco are influenced by the actions of qualified and competent pilots. Sunglasses, suntans, dress, and leisure pursuits can all be influential on trainees.

Typically, different people occupy the role of model, sometimes in a conflicting way. Thus a young student pilot, while trying to adopt many of the characteristics of the chief pilot of the commuter line operating out of the training airport, is still under the influence of fashions and pop heroes. There are some obvious conflicts for young student pilots who have peer group norms, attitudes, and values on the one hand (especially relating to dress, language, and behavior), and the new discipline of flying and pilot models on the other. This same conflict extends to the agricultural pilot talking of dodging power lines and treetops to dump a load of fertilizer, and the tight air navigation regulations restricting flight maneuvers for the young pilot. Given, however, that nonverbal behavior is more easily modeled than verbal, instructors have a psychological as well as a moral responsibility to practice what they preach. Actions model louder than words, in an aircraft or on the ground.

ACHIEVEMENT MOTIVATION

Achievement motivation is part intrinsic and part extrinsic. It consists of two major motives—the need to achieve success and the need to avoid failure. In some students the motive to achieve success can be higher than the motive to avoid failure. They are called *need-achievers.* For them, the greatest glory comes when the odds are 50-50—the odds do not favor them, but are not necessarily stacked against them. If the probability of success is high, it's a pushover. On the other hand, if the probability of success is low, around 10 percent for example, then there's not much point in wasting time on it.

Low need-achievers are more motivated to avoid failure than to achieve success. The relationship to persistence and probability of success is exactly the reverse of that for need-achievers. Those who fear failure will happily blow their chance of winning, as long as they preserve face. The 50-50 situation is thus the most threatening, not the most appealing. When the fear of failure is paramount, it's better to either win cheaply or to fail gloriously when odds are hopeless.

Some students will thrive on competition. Others adopt any tactic to avoid it. High and low need-achievers contrast in their persistence at key stages in flight instruction, such as final preparation for solo or instrument flight. High need-achievers behave logically, adjusting their aspirations upward in accord with prior success. They are ready for a more difficult challenge after each success.

Those who fear competition, however, tend to give up after a degree of success. Sometimes they persist after failure and continue to fail. For flight instructors, there is special significance in student persistence, especially in consideration of problem students who are not achieving as they should. The relative contributions of the student's need to achieve success, and the need to avoid failure, are crucial determinants of the instructor's response. These needs are the key to motivation and to successful flight instruction.

Intrinsic Motivation

Students become self-motivated when they are provoked to apply their abilities to a challenging but interesting task. Such self- or intrinsic motivation means that the involvement is of high quality; it is associated with strong, positive feelings; and it will be self-maintaining.

Obviously, intrinsic motivation is a highly desirable state for learning. How can flight instructors create the "challenging but interesting" situation?

CREATING OPTIMAL CONFLICT

The challenge that leads to intrinsic motivation of a student can come from a variety of instructional techniques. The challenge has to be large enough to spark interest and attainable enough to preclude avoidance. There are four methods a flight instructor can use.

1. *Surprise*: "What will happen if we use opposite rudder with aileron?"
2. *Perplexity*: "What circumstances can result in an altimeter having the same reading, yet be in three extremely different altitudes above sea level?"
3. *Bafflement*: "Given all the variables involved, how can one quickly calculate landing and takeoff distances required?"
4. *Contradiction*: "You're now automatically carrying out the prelanding checks on the downwind leg of the traffic pattern. However, there's one occasion when you won't carry out those checks. When?"

All of these strategies place the student in a slightly difficult situation involving a degree of conflict between what is known and what is being learned. Depending upon the personality of the student, that critical level of mismatch will differ widely. But when the right amount of mismatch occurs for that student, he or she will become intrinsically motivated.

Apart from creating optimal conflict, there are other ways an instructor can facilitate intrinsic motivation of the student.

Variability or surprise in the instructor's presentation can gain the student's attention and interest. One difficulty exists with the early courses for flight instructors related to the audiovisual component. Prior to the course, traditional practice in preflight briefs was for a standardized chalkboard summary to be provided. After the course, instructors simply switched to overhead transparencies instead of the chalkboard. They continued to fail to vary their methods, simply using the overhead like a chalkboard. They missed the point. The overhead projector, like the chalkboard, is an aid that extends the range of teaching media. The aim is to vary the approach to suit the student, the subject, the time, and so on.

Relevant background knowledge is vitally important for intrinsic motivation to occur. The student needs a backdrop of information for there to be effective mismatch. Background knowledge for future lessons can be introduced earlier. For example, in a consideration of the need to balance forces for straight and level flight, some allusions could be made to the symptoms that indicate an imminent stall, and the reasons for such an occurrence.

The learner's own pace is optimal for intrinsic motivation. It is extremely difficult to follow this principle when operating under the time and cost constraints of flight instruction. However, intrinsically motivated learning must be self-paced.

Finally, the student has to want to be involved. The persistent weekend flier who has saved for lessons is obviously no problem, but those whose parents have enrolled them for flight instruction may be quite another matter. The parents are simply sponsors of an uncommitted student. How can we get students to want to learn?

WANTING TO WANT TO LEARN

The learner usually cannot become intrinsically motivated until involved in the task. There is, therefore, a two stage problem: (1) starting the students off; and (2) maintaining persistence until intrinsic motivation takes over (assuming that it will).

Thus, it is important in flying instruction that the student actually fly before undertaking sustained study for theory examinations. Some students will be sufficiently interested to maintain motivation throughout the nonflying period, but others will be relying entirely upon extrinsic motivators. To assist the student to get to the stage when intrinsic motivation takes over, the instructor has a number of possible forms of action, mostly based upon our discussions so far in this chapter.

CONDITIONS AFFECTING INTRINSIC MOTIVATION

There are several situations that promote intrinsic motivation. They include *positive associations* of pleasure with flight; *social reinforcement* (instructor praise) and the example of admired figures (instructors and pilots); *freedom to make own choice* of activity and proceed at own pace; and *reward* when the activity was disliked to start with (unexpectedly and subtly—so there is not assumption of a "deal" or that the person is being "bought"). If the task is disliked, and the reward is unexpected, it may produce a change of attitude.

There are also situations that depress intrinsic motivation. These include *negative reinforcement* (threats of punishment); *surveillance* (supervising the learner too heavily and obviously); *unpleasant association* with flight or instruction; and *reward,* when expected, when task is already well liked, contingent on quality of performance (especially if failure is possible), unrelated to nature of task, and at a rate decided by an outsider. Great Aunt Nellie's promise of a new Porsche when Willie scores his commercial license does absolutely nothing for intrinsic motivation! Reward is results oriented, while intrinsic motivation is process oriented.

These two sets of conditions associated with intrinsic motivation give greater responsibility to the learner and bring a positive feeling about competence. It is not the reward that is important, it's how individuals perceive themselves in the context of the reward. Low intrinsic motivation brings a picture of a learner's poor perception of personal abilities, and an outsider "calling the shots."

These points come together when we consider the self-concepts of the students, that is, how they see themselves as budding pilots.

SELF-EFFICACY

This chapter started with reference to Fig. 5.1, which showed a continuum of sources of motivation from extrinsic to intrinsic. Another way to consider these extremes is by the use of the terms *pawn* and *origin.*

These two terms come from the work of psychologist Richard De Charms, who sees an *origin* as an intrinsically motivated person whose control is internal. The *pawn,* on the other hand, sees that his or her behavior is directed from outside himself or herself, by others who are more powerful. In De Charms's view, students will do only as well as the image or concept they have of themselves. In terms of command judgment this seems an important consideration for pilots in training.

To change a student's self-esteem so that progress in instruction will be improved is no easy matter. After a student has consistently received clear messages from flight records (not termed "hate sheets" for nothing!) he or she concludes that, "I am a failure—I cannot fly the aircraft the way they expect." With this information firmly embedded in his or her self-concept, the odds are that underachievement will continue.

One solution is for the instructor to prove them wrong—if time and cost permit. Make them succeed. Go right back to the first sequence and through graded exercises lead the students back to success.

Another solution is to provide students with the opportunity to make decisions themselves, rather than seeing themselves as being swept along in a program that spits out pilots at the end. This approach is associated with self-efficacy.

A student approaching an exercise, such as a forced landing without power or a solo navigation exercise, may form two sets of expectations: (1) "How effectively will I be able to carry out the exercise?" and (2) "What will the probable result be if I carry out the exercise?"

Students know they have to complete the exercise successfully to gain a license, and that the quality of performance will involve the opinions instructors and students hold of them. Efficacy determines the effort and persistence the student will put into the exercise. Efficacy expectations come from four sources of information: accomplishment, example, verbal persuasion, and physiological messages.

Accomplishment

Having successfully completed a solo landing, the student has increased confidence about repeating that success. If the first approach or hold-off was seriously misjudged, perhaps resulting in an undershoot or overshoot situation, there will be a negative effect on intrinsic motivation. It is important to see that motivation is linked with belief of success, not success itself as seen by others. Thus a hypercritical instructor can lead students to believe they are not cut out for flying (causing them to give up), while an understanding instructor would lead the same students to a more optimistic assessment.

Example

Students see their peers, other pilots, and instructors carry out numerous landings in the course of a day at the airport. It's logical enough for the student to conclude, "If they can do it, I can." This expectation is not as powerful as the first accomplishment, however. It's possible that the student could conclude "but I am not as clever as they are."

Verbal Persuasion

This, too, can work both ways. "Of course you can do it," casually muttered by an instructor looking out the window as he sets up the aircraft for a glide approach, is hardly convincing for the student. But verbal messages that agree with nonverbal communications can persuade.

For example, an instructor could give the thumbs up sign of approval at the end of a series of gliding turns, then say something like, "Okay, Marion, take her up again. Your gliding turns were balanced and accurate, you remembered the lookout and to clear the motor through the descent, plus the use of carburetor heat. Let's put it all together now and make a glide approach to landing, as we discussed in the briefing room." The confidence in the student's success is implicit in the instructor's confidence and matter-of-fact acceptance that it is the next stage for which the student is prepared. Instructor actions, such as an almost panic grab at the control column midexercise, speak much louder than the most carefully chosen words.

Physiological Messages

If a student's arousal system is working overtime (say, before a flight check), the associated expectation of the student is for failure on the test. However, if the student goes into the test without unpleasant physical arousal, this will contribute to a feeling of efficacy. Efficacy expectations vary in strength and generality, and these four sources can affect each other. Students may realize that they landed quite successfully on the last occasion, but could also be experiencing an unprecedented nervousness on this occasion. The cross-wind component, some wind shear, some turbulence on final, and the student thinks, "That was OK last time, but this isn't quite the same." This leads us to consider the generality of expectations of success. Students who failed their first test on the principles of flight may or may not believe that they will also fail the meteorology test. Down-putting encountered in one exercise can generalize to others, with an effect on the student's learning and self-concept.

It need not be that way. The student can feel good about successes as a football player, pool player, athlete, or speller and accept weaknesses in mathematics, chess, or table tennis. Beliefs about self-efficacy are closely tied to particular tasks.

Attribution Theory

In the formation of their efficacy beliefs, it is important that students attribute success or failure to causes that will motivate future performance rather than causes discouraging further involvement. A student's belief in the slowness of his or her progress in navigation exercises is derived from the instructor's comments implying that the student is not achieving at the average rate. As we saw earlier, the effect will be to reduce or remove the student's intrinsic motivation for future navigation exercises.

This statement needs qualifying, however. It depends upon what the student attributes the lack of success to. If the student attributes failure to something that isn't likely to occur again, or that can be prevented from recurring, then the initial lack of success may make little difference. For example the student may think, "Doppler is a rotten instructor. Just my luck to end up with him. All the trainees with Smiley have gone solo. If I can switch instructors I'll be OK." or "I just didn't put enough time into my flight planning. Next time I'll be more thorough and spend more time on it."

Alternatively, the student can attribute a failure to internal factors. For example: "I just can't handle navigation. I haven't got what it takes." or "I failed the navigation because I couldn't handle the work load of flying the aircraft while navigating as well. After more flying experience, it should be different."

Thus two dimensions underly judgments about attribution: stable-unstable, and internal-external. This makes four possible kinds of attribution, and different reactions following success or failure. Let us first discuss these possibilities in relation to failure.

1. *Unstable-external.* The failure is attributed to bad luck (in getting Instructor Doppler). Further bad luck is unlikely and the student feels little shame in the failure. Thus, the likely outcome is indeterminate with respect to future motivations.

2. *Unstable-internal.* The failure is attributed to insufficient effort. The student can retain belief in his or her basic ability, but there is shame in not trying. However, if the student is prepared to work harder, he or she can be optimistic in the future.

3. *Stable-external.* The failure is attributed to the fact that navigation is recognized as a difficult phase of learning to fly. There is little shame but little expectation of improvement. This attribution is unlikely to result in subsequent intrinsic motivation.

4. *Stable-internal.* The failure here is attributed to an internal factor (low ability) that is unlikely to change. The instructor commented, "If you can't plot an alternate destination you can't get your license—and everyone else in the flight has passed the exercise."

The most damaging possible attribution is to low ability. Instructors should therefore avoid contexts that lead the student to this attribution—criticism, sarcasm, comparison with others after failure. Attributing failure to stable factors decreases future expectations of success and hence motivation.

So far we have looked at attributions following failure. But what about those following success?

1. *Unstable-external.* Success is attributed to luck, which could turn in the future. So failure is as likely as success next time. Pride in the accomplishment is minimized. Intrinsic motivation is likely to be low.

2. *Unstable-internal.* Success is attributed to extra hard effort. "I really worked for that exam, but I don't know if I can put that much into it in the future." There is more pride in success than in (1), but there is also the implication that the performance won't be sustained. Future history will decide the motivational outcome.

3. *Stable-external.* Success is attributed to an easy task. "Incipient spin

recoveries are a pushover. Dead easy." There is an expectation of future success but little pride is involved, and little likelihood of intrinsic motivation.

4. *Stable-internal.* Success is attributed to ability. "I'm good at this. Flying is my thing!" Pride is felt, and there are expectations of future success. Intrinsic motivation is likely to be high.

In each of the eight situations, did you notice the link between pride and intrinsic motivation? A student's sense of pride is worth cultivating: it's the price of intrinsic motivation.

WHY PEOPLE MAKE DIFFERENT ATTRIBUTIONS

There are three reasons people make the attributions they do.

1. Other people. Other people, especially authority figures, such as instructors or examiners, affect beliefs in one's ability. Thus instructors have a special responsibility to avoid linking results with "causes" that would lead to attributions adversely affecting motivation.

2. Situations. Flight instruction is normally regulated by rules and routines in order to operate efficiently. The disadvantage is that this places many restrictions on the individual's choice (of, say, dress, what to learn, when to learn it). Thus it is easier for stable-external attributions to be made ("What's the point . . . it's all up to them at any rate.") involving no personal responsibility.

3. Personality. Personality factors affect attributions. For example, females tend more than males to attribute success to luck rather than to ability, and to rate their ability lower. Individuals with low self-esteem tend to make internal attributions (low ability) following failure. High need-achievers attribute their failures either to external factors or to lack of effort (but not lack of ability), but attribute success to internal factors (ability and effort) to which greatest pride is attached.

Low need-achievers, on the other hand, attribute their failure to lack of ability and their success to luck or an easy task.

The general implications are clear. Belief in self-efficacy and appropriate attributions of success to ability and of failure to lack of effort are likely to motivate the individual to future achievement. Attributions of success to luck or to an easy task, and failure to lack of ability are likely to kill subsequent interest in the tasks that achieve this effect.

Summary and Conclusion

Methods of motivating students can affect the student's attributes to flying in both the short and long term. The methods can also affect the student's self-concept in a way that will either facilitate or obstruct the development of command confidence and ability.

Punishment is the least predictable instructor response. Alternatives include positive and negative reinforcement or the extinction process.

The performance of a student pilot in a flying exercise can be related to the individual's feelings. If the student feels confident, capable, and positive towards the stalling exercise, there is a strong probability that the exercise will be successfully completed.

Discussion Questions

1. From your experience, why do people wish to learn to fly an aircraft? Categorize your answers as either extrinsic or intrinsic motivators.
2. What do you see as the major implications for flight instructors?
3. Cite a way in which you could use optimum conflict to challenge the intrinsic motivation of a student pilot.
4. One of your *ab initio* student pilots is obviously underachieving. List some of the methods you would use to improve the student's self-esteem.
5. In your own words, explain the Premack principle and give an example of its application to flight instruction.
6. How does attribution theory explain success and failure in student pilots?

Further Reading

Biggs, J. B., and R. A. Telfer. *The Process of Learning*. Sydney: Prentice-Hall, 1987. Chapter 3 is on motivation, attribution theory, pawns, and origins, and Chapter 4 is on student approaches to learning (including reflective self-awareness and self-direction).

6 Evaluating Learning

THE FOLLOWING QUESTIONS are answered in this chapter: What is the difference between summative and formative evaluation? What is meant by "normalizing" test scores? How does qualitative testing differ from quantitative testing? When is a flight test reliable? When is a flight test valid?

After reading this chapter, you should be able to: (1) construct a norm-referenced or criterion-referenced test; (2) avoid bias and unreliability as an examiner; (3) be aware of the advantages and disadvantages of different evaluation methods for particular aspects of flight instruction; and (4) provide alternatives to the usual pass/fail or percentage means of communicating student results.

Some Basic Concepts in Evaluation

Learning and evaluation are closely interdependent, but their relationship is not a simple one. On the one hand, learning and evaluation are two sides of the same coin; evaluation is necessary to determine the course of future learning or the remediation of previous learning. On the other hand, evaluation and learning are sometimes in conflict: evaluation can stifle learning (where there is test-anxiety). Evaluation is an important topic in the context of learning about flight and it has its own concepts, skills, and techniques.

In this section, we take a nontechnical look at some of the assumptions and distinctions involved in the relationship between evaluation and learning.

TESTS

A test is an instrument for measuring educational outcomes. It may consist of one or several items, and it may be in one of several forms (multiple-choice, essay, practical, and so on). A test may yield a numerical score or a qualitative category, and the instructor has then to make a reasonable decision on the basis of this information. The use of test information is based on the kinds of distinctions considered below.

In flying, tests are usually used to measure abilities or achievement. Cognitive tests, with which this chapter is mainly concerned, can be either instruction-dependent or instruction-free. *Instruction-free tests* are so called because they measure performances that are relatively free from the effects of instruction: they are ability tests that characterize the person, and include measures of general abilities such as IQ. *Instruction-dependent tests* evaluate how well students have learned particular bodies of knowledge or skills that have been the subject of instruction. These are called *attainment* or *achievement tests,* and are usually constructed by the instructor. We are chiefly concerned with attainment tests in this chapter.

FORMATIVE AND SUMMATIVE EVALUATION

A distinction can be made between formative and summative evaluation. Formative evaluation provides feedback to both instructor and learner and is used during the teaching process; summative evaluation indicates how well material has been learned after teaching is completed.

Formative evaluation tells the student and instructor how well the course is being learned at any given stage in the learner's development. Formative evaluation is intended simply to give information to the participants, not to punish, grade, inform the public, or award scholarships. Summative evaluation tells the student, instructor, and client how well the course has been completed. Thus, formative evaluation is continuous, diagnostic, and remedial, while summative is terminal, finite, and descriptive.

As techniques of formative and summative evaluation are quite different, instructors must be clear as to what they are trying to do when administering a test. Is it to find out how well students are taking to the material? Is it to find out if there are any misunderstandings so far—if the teaching is aimed at the right level? Or is it to gain information for grading purposes? Will the test data appear on final reports? The answers to these questions point to the different tests needed.

NORM-REFERENCED EVALUATION

Norm-referenced measures are interpreted according to the performance of an individual in relation to others. The meaning of the score used, such as rank, standard score, or percentile, depends on this relationship. For example, "Joe was top of the flight this year" is a norm-referenced statement of Joe's ability. It tells us nothing about what Joe actually did, only that whatever he did was better than anyone else. Here the standard shifts according to the performance of the class as a whole.

Because norm-referenced techniques are widely used in educational measurement, it is important to look at the more common of these in some detail. Probably the most common of all is *ranking*. Joe was top of the flight. Ann and Gary tied for second, and Julie was third. The criterion here is who is better than whom, where "better than" is defined by the context. Other norm-referenced measures, based on ranking, are percentiles, the median, deciles, and quartiles.

The Normal Curve

Some norm-referenced measurements are based on the assumption that what is being measured is distributed along the normal curve.

The normal curve is symmetrical and bell-shaped. (See Fig. 6.1.) It is obtained whenever a characteristic is randomly distributed throughout a population. Take the case of intelligence. Because it is mathematically convenient, IQ test constructors assume that intelligence is normally distributed: the average or mean is then set at 100. We now move from this average in either direction and mark off equal units, which are called standard deviations (SDs). IQ tests are usually constructed to have a standard deviation of 15 and a mean of 100. The relationship between IQ, SD, and percentage of people occurring at a given point in the curve is given in Fig. 6.1.

Since each SD step is 15 IQ units (the test was made that way), one step either side of the mean encompasses an IQ range of 85 to 115 (i.e., 100 ± 15). This includes the middle 68 percent of all cases. Another step in either direction gives a range of 70 to 130, which includes 96 percent of all cases. In other words, in a sample of 100 randomly selected people, one would expect to find only 4 people outside this range: 2 with an IQ below 70, and 2 with an IQ above 130. If one takes a further step, the range becomes 55 to 145, which includes all but 2 persons in 100.

This is not only true of IQ scores, but also of ability tests in general. The distribution of scores on any test can be normalized by setting the mean score at 100 (or any other arbitrary value), and rescaling all scores above and below 100 according to the distance of each score from the mean.

Advantages of Normalizing Scores

There are three advantages of normalizing scores.

1. The results are easy to interpret. As soon as we know a person's score, and the mean and standard deviation of the test distribution, we can tell exactly how a person stands in relation to others.

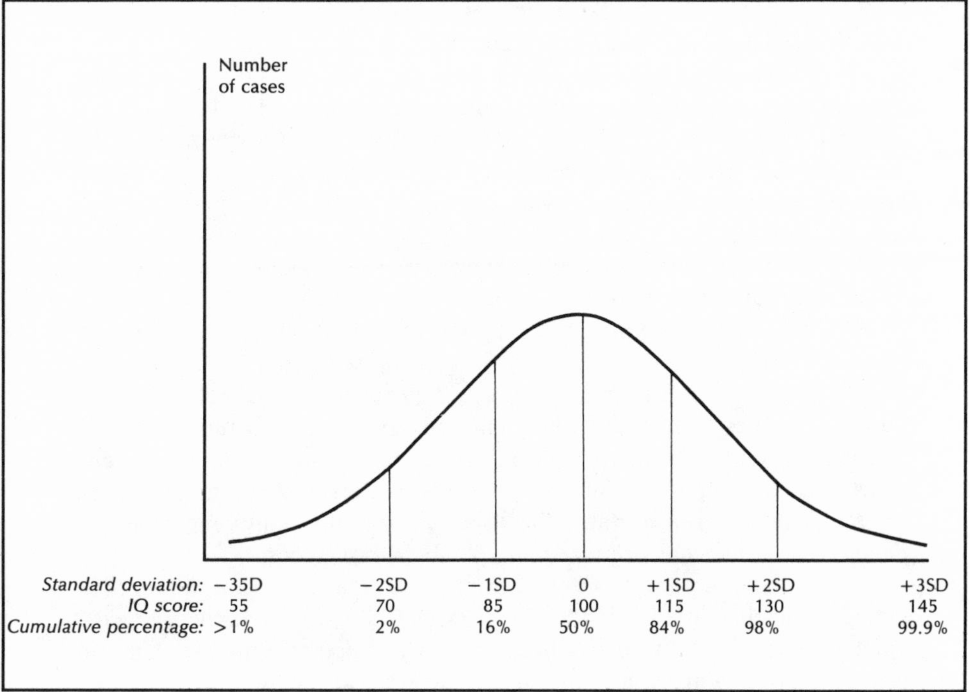

FIG. 6.1. **The Normal Curve**

2. Results may be compared between tests. If John receives 20 marks out of 25 in navigation and 33 out of 40 in meteorology, it is difficult to tell how he stands in one subject as opposed to another. When we convert to percentages (80 percent in navigation and 82.5 percent in meteorology), it seems that meteorology is marginally better than navigation. But what if that 80 percent were the highest navigation mark in the class, and there were four other students getting higher than 82.5 percent in meteorology? Relative to the class, navigation is now John's best subject, not meteorology. To know exactly how each person stands in each subject, we would normalize the grades using the scores of each class as a standard.

3. Results from different tests may be combined. Normalizing expresses the mark on different tests in terms of equivalent units. If we wanted to add John's marks in navigation and meteorology to obtain a composite achievement score, then normalizing is the correct procedure. It would obviously be misleading to add the original raw scores of 20 and 33; and for the reason mentioned above, even adding percentages has its problems.

Disadvantages of Normalizing Scores

Normalizing scores has four disadvantages.

1. The normal curve is a mathematical model. It describes a perfect case that may or may not correspond to reality. It does correspond fairly well when the characteristic being measured occurs randomly, but when nonrandom constraints distort this symmetry, the normal curve becomes less appropriate.
2. Norm-referenced evaluation emphasizes competition.
3. Normalization makes no sense in small classes.
4. Normalizing requires extra work on the part of the evaluator.

To summarize, then, normalizing is a useful procedure only when the number of cases (students) is large enough; the test scores reflect reliably and validly some characteristic that is reasonably supposed to be normally distributed in the population; certain kinds of statistical techniques are to be used; and most important, the situation logically requires normalized data. These conditions are rarely met in flight instruction.

Norm-referenced tests are used primarily to make decisions about individuals, and their use is obvious in "rationed" situations. If there are 100 students competing for one scholarship, it is important to use a norm-referenced test to make sure the best student is awarded the prize.

CRITERION-REFERENCED EVALUATION

Criterion-referenced measures are used to make a different sort of decision about individuals: it concerns particular task requirements only (that is, whether or not an objective has been met). Can this person (irrespective of how others perform) carry out an emergency landing or fly by instruments? An example of such a test appears (excerpted) in Insert 6.1. It was provided by the Australian Department of Aviation.

Criterion-referenced tests are those that "yield measurements that are directly interpretable in terms of specified performance standards." This means that the individual is evaluated in terms of some prescribed standard irrespective of the performance of other individuals. Criterion-referenced measurement focuses attention on whether or not the learner has learned what was intended to be learned. If instructors decide that a certain objective is important, then presumably their major concern is with whether or not a particular student has achieved it, rather than how much of it was achieved relative to others.

A common example of a criterion-referenced test is a flying test as shown in Insert 6.2. In order to pass the test, one has typically to demonstrate (a) verbal knowledge of flight legislation, and (b) pass-fail prowess in

INSERT 6.1. Criterion-referenced

Test Flight Proficiency Test—Class One Instrument Rating

An applicant for a class one instrument rating shall demonstrate in flight (except where otherwise specified) his ability to perform solely by reference to instruments the undermentioned flight maneuvers and the radio aids procedures applicable to the type of radio aid for which a rating is desired. Unless otherwise stated these maneuvers and procedures shall be performed having recourse to all available instruments.

General Flying

A flight check on an air route embracing:

(a) flight planning and navigation;
(b) preparation for flight;
(c) airways operating procedures;
(d) all maneuvers associated with the normal operation of the airplane type.

FLIGHT MANEUVERS AND DEGREE OF ACCURACY:

1—Ascent: Heading ± 5°.
Airspeed ± 5 knots.
2—Ascending Turns: Airspeed ± 5 knots.
3—Cruise: Heading ± 5°.
Height [Altitude] ± 100 feet.
4—Steep Turns of Approximately 40°
(Initial issue only): Angle of bank ± 5°.
Height [Altitude] ± 150 feet.
5—Descent and Descending Turns: Airspeed ± 10 knots.
Rate ± 20 per cent of designated rate.

Actual rate of descent wll be calculated by comparison of time and change of altitude.

6—Asymmetric Flight (or reduced thrust in the case of center line thrust aircraft)

The applicant shall demonstrate his competency in the following:

(a) **Take-off:** Attaining optimum performance following simulated failure of an engine at least 10 knots. above take take-off safety speed and either the under-carriage fully retracted or the aircraft has attained a height of at least 50 feet above terrain.
Heading ± 20° initially, then ± 5°.
Altitude continued climb.
Airspeed not lower than take-off safety speed.

Note: Simulated instrument flight should be introduced not later than that point at which a pilot would normally lose visual reference during a take-off at night, or because of the nose high attitude of the airplane on initial climb.

(b) **Cruise:** Heading ± 15° initially, then ± 5°.
Height [Altitude] ± 100 feet.

(c) **Turns:** Left and Right: Height [Altitude] ± 100 feet.

(d) **Baulked approach:** Failure of an engine is to be simulated prior to or during an instrument approach using an aid for which the pilot is endorsed and an asymmetric baulked approach from the approach configuration is to be carried out from the minimum altitude in accordance with the missed approach procedure for that aid.

Note: Correct handling of ancillary controls and adherence to engine limitations as applicable is a requirement for satisfactory performance of these maneuvers.

INSERT 6.2. **Qualitative Testing: Second Class License Renewal Base, Check Exercises, F28 Aircraft**

SEQUENCE:

1. PRE-FLIGHT BRIEFING.
2. PERFORMANCE QUIZ—MINIMUM FOUR QUESTIONS.
3. EMERGENCY QUIZ—PHASE 1 and "A" DRILLS.
4. START—NO AC, CROSS BLEED AND/OR STARTER VALVE LIGHT REMAINS ON.
5. FIRE TAXYING.
6. VIS TAKE-OFF—FAILURE—VMC SEGMENTS TO DOWNWIND (MIN. 1000′ AGL DAY OR DAY/NIGHT CIRC. IF HIGHER. MAX. 1500′ AGL).
7. SINGLE ENGINE CIRCUIT AND LANDING.
8. BRAKE FIRE.
9. I/F TAKE-OFF (NO S/C)—FAILURE—IMC SEGMENTS TO 1500′ AGL.
10. RESTORE POWER TO BOTH ENGINES, ESTABLISH CRUISE AT NOMINATED AIRSPEED AND HEADING, ENGINE FIRE IN CRUISE.
11. JOIN CIRCUIT, SINGLE ENGINE MIN. WEATHER/MIN. VIS CIRCUIT.
12. SINGLE ENGINE MISSED APPROACH.
13. SINGLE ENGINE CIRCUIT AND LANDING.
14. BRAKE FIRE.
15. DE-BRIEFING

NOTE:

If time permits, alter the sequence as follows:

(a) Item 7, conduct a SINGLE ENGINE MISSED APPROACH TO 2000 FT.

(b) Single Engine Min. Weather/Min. Vis Circuit and Landing.

(c) Item 10, conduct in 12 VOR HOLDING pattern at 4500–6500 ft. Complete Drills and Pre-Landing Checks.

(d) Item 11, a 12 VOR APPROACH, SINGLE ENGINE, RAW DATA.

(e) Item 12, Missed Approach from the MINIMUM ALTITUDE.

(f) Item 13, Single Engine Circuit and Landing.

Courtesy East-West Airlines (Operations) Ltd., Australia.

a test flight. If everyone (or nobody) passed the test on the first go, then no doubt questions might be asked about the suitability of the standards, but that is a separate issue.

The question of which form of summative evaluation to use—norm- or criterion-referenced—will depend upon the kind of decision that is to be made. Where instruction is for the purpose of learning essentials, with variable time allowed, criterion-referenced testing is preferable. Where, however, it is desirable to encourage maximum development, and where time is constant, norm-referenced testing is best.

TESTING: QUANTITATIVE EVALUATION

In evaluating a learning episode, two basic questions may be asked: *how much* has been learned; and *how well* has it been learned. The first is a quantitative question, the second qualitative.

Quantitative evaluation is comparatively easy to carry out. The instructor wants to know how many checks the student can recall, the number of problems he or she can solve correctly, the number of rules she or he can apply, and so on. There is an elaborate technology to help instructors evaluate quantitatively.

The dominance of quantitative evaluation is evident when we look at the use of multiple-choice tests, which are based on the number of items correctly answered. Even essay-marking has a quantitative bias in practice. The almost universal procedure in marking open-ended essay responses is to award a mark for each relevant point made and convert the ratio of actual marks to possible marks into some kind of number. The instructor then adjusts the final mark for overall quality, so that a final grade is arrived at. While everyone would agree that quality is important, that final adjustment for quality is usually quite subjective. These terms are rarely spelled out for the student, so that the student can understand and benefit from them. Their effect on the final grade is also by means of an equally private analysis. Even in essay-marking, then, "how much" tends to call the grading tune more than does "how well."

Test Reliability and Validity

Whatever the purpose and nature of tests, they have two general properties that teachers need to understand: reliability and validity.

RELIABILITY

An important property of a test is its stability or *reliability.* The test, like any measuring instrument, should perform identically from day to day, irrespective of who is administering it. A clock that is unpredictably running fast and then slow is obviously unreliable, so is an elastic tape measure. Thus we have to assume any test we administer renders a reliable estimate of the student's level of achievement. Variations in the test score must be assumed to be due to variations in the characteristic being measured, rather than to unrelated factors that therefore become errors of measurement. There is always some error of measurement present, but this should be as small as possible.

Sources of Measurement Error

There are three main sources of error.

1. *In the testing environment.* If the test is administered in a noisy room with many distractions, the student's performance is likely to be erratic. There may be a tendency to misinterpret or skip items or make mistakes. Another factor in the testing environment is the way the test is presented. Are instructions given clearly? Is the manner of presentation likely to arouse or reduce anxiety? If the test is timed, is it timed accurately? Standardized administration is particularly important in norm-referenced tests (e.g., IQ tests), because the basis of norm-referencing is comparison (of individual with group norms). In criterion-referenced tests, however, the test conditions should be optional for each student, so that the best performance is extracted. Thus, criterion-referenced tests may quite reasonably be administered under differing conditions: some students might need a longer time than others; some will do well under pressure, others under relaxed conditions. Instructors need to be very careful about whether they standardize or optimize the test of environment.

2. *In the learner.* The reliability of the test will also be affected by changing factors in the learner. These might include tiredness or illness on the day of the test, poor reading skills so that test items are capriciously misread or misunderstood, feelings of anxiety, and so on.

3. *In the test itself.* A most important source of unreliability is in the test itself. If items are poorly or ambiguously worded, they may be interpreted differently by different people; or one person may interpret the same item differently on different occasions. Likewise, the scoring of the test must be consistent and free of bias, so that the same scores are obtained by different competent markers.

Measures of Reliability

Unreliability thus has several sources: in the testing situation, in the learner, and in the test itself. How do we know whether a test gives a reliable score or not?

1. Test-retest reliability. If one person under identical conditions on the same test achieves the same score on two separate occasions, the score is said to be reliable or stable. Test-retest reliability makes good theoretical sense, but there are two practical reasons that may make it unsatisfactory. First, practice may have an effect: having done the test once, the student will find it easier to do next time, and hence will get a higher score. Second, the student might actually have changed in the interim, having learned or forgotten something that affects the score. The test may reflect these changes accurately, but it would appear to be unstable.

2. Split-half reliability. One way of beating these problems is to divide the test into two halves: the odd items and the even items are scored separately, and the two scores are correlated. If the correlation is high, the two half-tests are yielding compatible scores.

3. Internal consistency. A test is internally consistent if all items intercorrelate: whatever they are measuring, at least they are consistent.

4. Interjudge agreement. A test must be "marker-proof." Two instructors marking an exam must be able to agree, within limits, about the marks awarded to a student's responses. Equally, the same person must be able to make the same judgment about a response on two different occasions. If an examiner cannot determine which way a response is to be marked, either the examiner is incompetent or the test is a poor one. Since we assume that all examiners are competent, we would have to say, therefore, that the test is unreliable and of little value.

VALIDITY

The validity of a test is its ability to measure what it is supposed to measure. This is the most important property of a test. There are several kinds of validity.

1. Face validity. Competent judges would agree that the test appears to be measuring what it is intended to measure. Particularly with attainment and ability tests, this agreement is fairly easily obtained. This form of validity is a good start, but is not in itself sufficient to establish validity.

2. Content validity. Test scores should relate to other measures, the validity of which is known or assumed.

3. Predictive validity. The test should enable us to predict future performance. Predictive validity is particularly important when the test is used for selection purposes. It must be demonstrable that those selected actually do perform better than those rejected.

The proper determination of validity requires experimental work, statistical know-how, and facilities that the instructor will not have. The only kind of validity check readily available is that of face validity, where an instructor uses personal judgment and possibly that of a colleague.

In this section, we have examined some basic concepts and distinctions in evaluating learning. The important underlying issue is that there is no simple relationship between learning and evaluation. As we have seen, there are different kinds and purposes of learning in aviation, and the same evaluational strategy will not do for each.

Methods of Testing

Ground instructors are continually evaluating and testing students in an informal way—observing, questioning, getting them to question each other, asking students to give talks or displays, correcting work, and so on. The methods of formal testing are more limited, and it is with the more important of these that we are concerned in this section.

THE ESSAY

The *essay* may be defined as a continuous piece of prose written in response to a question or problem. It appears to be a simple and direct way of finding out what a student knows and how well the student can think about a topic. A question is asked, and the pupil is then given what should be sufficient time to organize thoughts and express them succinctly.

There are many variations of the essay technique: the *timed* examination, with students having no prior knowledge of the question; the *open-book* examination, where students usually have some prior knowledge and are allowed to bring reference material into the examination room; the *untimed* examination, where students can take as long as they like, within reason; the *take-home* examination, where students are given notice of the questions and they are given several days to prepare answers in their own time; and the *assignment,* which is an extended version of the take-home and which comprises the most common of all methods of evaluating by essay.

To avoid unreliability it is very important that markers become aware of the criteria they are using during marking. Markers are also swayed by their own prejudices in awarding marks for quality. Students will thus get higher grades than they deserve if they correctly divine what the marker wants to read.

Further, personal characteristics of the markers can influence judgment. For example, markers who are untidy writers mark neat and untidy essays equally; neat writers, however, tend to downgrade untidy essays heavily. These findings suggest that, unless all essays are typewritten, only instructors with untidy writing should be permitted to mark them!

Personal preference for appearance or personality can cause a marker to grade one student higher than another. For example, it has been demonstrated that attractive female students tend to receive higher grades from male teachers than unattractive ones (we could find, as evidence of the converse possibility, that attractive male students are at a corresponding advantage with female teachers).

Not only may different markers vary greatly, but the same marker can also vary from script to script. Apart from momentary lapses, fatigue, and so on, another systematic source of unreliability is the order effect. When a marker sits down to a batch of thirty or so essays (the more there are, the worse the effect), the first half-dozen tend to set the standard for the next half-dozen, which in turn reset the standard for the next. Also, a moderately good essay following a run of poor ones tends to be marked higher than it deserves; similarly, if it follows a run of very good ones, it would be marked down. The marker's standards thus tend to slide up or down according to the quality of the essays being marked at the time.

Reactions to findings such as these have been in two directions. One is that the essay medium is so unreliable that it should be replaced by objective tests. This implies, however, that students would be given less and less opportunity to carry out writing in continuous prose, when they are already given too little. The other is that, despite its obvious limitations, the essay is a useful technique and it would be worthwhile improving the faults of essay-marking.

The following are some precautions that may help mitigate the order effect:

1. *All marking should be done "blind"* (that is, the marker is unaware of the identity of the student, and is thus less likely to be influenced by irrelevant factors such as appearance or personality). Blind marking makes it less possible for instructor expectations to become a self-fulfilling prophecy.

2. *"Blind" rechecking* (with the original mark concealed, to ensure that standards are remaining fairly constant).

3. *Decide what criteria are to be used in marking.*

4. *Grade generally at first,* say into "excellent," "pass," and "fail," and later try to discriminate more finely within these categories. All borderline cases should be reviewed or rereviewed.

5. *Spot-checks,* with particular care for borderline cases, should be carried out by an independent marker. Gross disagreements should be resolved between the markers or by recourse to a third marker.

6. *The wording of the original questions should be checked* for ambiguities by an independent marker.

7. *Guard against bias due to handwriting* by first quickly scanning all

papers and deciding whether there are any particular ones that need to be rewritten.

8. *Each question should be marked across all students,* if several questions per student are involved (for example, where an exam booklet is used). This procedure sets a standard for each question, and prevents "halo effects" from question to question. Between questions, the papers should be shuffled to prevent systematic order effects.

9. *Use a "model answer,"* with points awarded for each congruence between the model and the student's essay. This is particularly important where essays are sectioned off to different markers. On the other hand, this procedure does away with one of the major advantages of the essay, the fact that the student can structure the response. Clearly, the unthinking application of the model-answer technique would penalize the highly original student.

THE OBJECTIVE TEST

The objective test is "objective" only in the sense that the scoring and collation of marks is, in principle, independent of the prejudices and judgment of the scorer: indeed, scoring is now often done by machine. However, the prejudices of the test-constructor can very easily show themselves in the alternatives chosen and in the ones designated as correct.

Two main forms of the objective test are in common use: where two alternatives are provided (the true-false test) and where several—usually four or five—alternatives are provided (the multiple-choice test). Other versions involve matching procedures (colors of fuel with octane rating, cloud types with names, etc.), filling in blank diagrams, completing sentences, and so on. Of all these techniques, the multiple-choice format is the most widely used and the most acceptable. True-false is open to the obvious objection that a score of 50 percent could be obtained by guessing, although it is possible to offset this by penalizing wrong replies. We shall mainly be concerned with the multiple-choice format here.

THE SHORT-ANSWER TEST

A test format that falls between the objective and the essay is the short-answer technique, where essay-type questions are set, and the student is asked to answer in note form, using abbreviations and avoiding elaboration. This format is useful for getting at fairly factual and straightforward material particularly for addressing or interpreting diagrams, charts, and tables.

The usual presupposition behind the short-answer test is that the examiner is after something quite specific, and there is only one correct answer. The use of the model-answer technique is more justifiable here than in the essay proper. The short answer is thus well suited to criterion-referenced testing where the answer is too complex (or otherwise inappropriate) to put into a standard multiple-choice format. Another advantage of the short answer is that it is less susceptible to test-taking strategies than the ordinary multiple choice. In the latter, it is possible to give a correct response by a process of elimination (the student does not know enough to decide that a response is correct, but does know enough to realize that three of the four alternatives are incorrect). Further, multiple choice depends upon a process of recognition, whereas short answer depends upon recall; and it is much easier to recognize something as familiar than it is to recall it. If the examiner felt that it was more appropriate to test recall rather than recognition, then the choice is the short-answer format.

PRACTICAL TESTS FOR STUDENTS AND INSTRUCTORS

Here the learner is put in a practical situation that is identical to the behavior that will be required when instruction has ceased. This is the only logical way of evaluating flight instruction.

Criterion-referenced evaluation is most appropriate for evaluating practical ability. The objectives should be quite clear-cut: the student has to perform certain behaviors to a specified standard. It should therefore be a simple matter, in most cases, to specify what these behaviors are, whether the learner passes, and if not, why not. Let us take, for example, a particular segment of instructor behavior: the junior instructor is adopting a lecturing style that requires that he tell the class a certain principle and then question the pupils. The desired behaviors are placed on a checklist: many behaviors might appear on such a list (audible speech, eye contact, use of visual aids), depending upon circumstances. The instructor should be well aware in advance of what such a list might contain, and should be given fairly immediate feedback, perhaps in conjunction with a videotaping (in which case, trainees would find it valuable to rate their own performance before discussing their supervisor's).

The diagnostic value of such a checklist is considerable, far more so than a typical norm-referenced evaluation of a practical teaching session, which might read:

A: Definitely superior; among the best in the year
B: Above average
C: Average
D: Below average, but meets minimal standards
E: Not up to standard

Clearly, such a rating is singularly uninformative. Even the "A" student doesn't know whether she or he is really a good instructor or merely the best of a bad bunch; while students of "B" gradings and below have no idea, in the absence of other information, what was really wrong with their performance. A criterion checklist, moreover, is likely to lead to more reliable measurement than an overall grading: two raters are more likely to agree on more specific, detailed behaviors (for example, a student's audibility or skill in question-handling) than on whether the student is "average" or "above average" overall. The critical point, with the checklist approach, is whether the items of listed behaviors include the most important aspects of teaching. Again, we come back to the matter of professional judgment.

One general advantage of the practical as a medium of evaluation is its apparent face validity: the student is being assessed in a situation that models reality (as Insert 6.1 shows). The closer the practical is to the real thing, the greater its validity. It is, however, often difficult to get psychologically close to actuality. The presence of the evaluator distorts the situation so that nervous trainees, in particular, are likely to behave quite differently from the way they would behave if they were not being observed. Nevertheless, the amount of distortion introduced by such factors is probably less than would occur if practical skills were assessed exclusively by written examinations (which are themselves likely to be affected by anxiety and other factors).

ORAL TESTS: PRESENTATION AND INTERVIEW

Oral forms of evaluation include the pre- and postflight briefs, and the *interview,* where the instructor questions the student on a one-to-one basis. The great advantage is that this is a two-way process. Skillful examiners can plumb the depths of students' knowledge and abilities in ways that they may not be able to prescribe in advance. They can also provide feedback to students, and thus serve a valuable formative function. On the other hand, the success of the technique depends to a large extent upon the mental chemistry of the participants: some such interviews go off with a bang, with enjoyment and stimulation to both parties; others can be painfully embarrassing. This instability is reflected in the extraordinarily low reliability of the interview, which is in fact lower than any other evaluation medium. It should be noted that this unreliability applies only to the open-ended interview; a structured interview that is conducted with a criterion-referenced checklist is not open-ended and may be quite reliable.

Constructing Tests

CONSTRUCTING A NORM-REFERENCED TEST

Say we are to construct a 50-item test with 4 choices per stem for a course in meteorology. There are 3 stages involved: (1) deciding what kind of items relate to the intentions of the instructor or examiner; (2) constructing a pool of potential items (three times the number required is a common rule of thumb) and testing these out by the process of item analysis (described below); and (3) deriving the final version of the test.

First, one defines the item types, and then collects examples of possible items. These should be checked for wording (ambiguities, clarity), preferably with a colleague's help, and then roughly 150 items should be administered to a few classes. The test papers are then scored and placed in rank order, with the highest scorer on top, the next second, and so on, with poorest last.

A criterion-referenced test might have subsections, with several items in each, dealing with "enabling" objectives that lead progressively towards the terminal objective. Instead of a final total score, the ideal test would provide a series of subscores so that one could tell at a glance a student's progress in relation to the final goal.

To some extent, instructors are on their own in defining mastery and judging validity. In deciding upon what mastery ought to mean in a given context, instructors must rely heavily upon their professional judgment. However, the process described by Robert Gagne (see "Whole Versus Part Methods," Chap. 3) provides a clear guide.

CONSTRUCTING CRITERION-REFERENCED TESTS

In norm-referenced tests items are selected as good or bad depending upon whether or not they discriminate the good from the poor students. Test items that discriminate most effectively are those that are answered correctly by about 50 percent of the students: the *best* 50 percent. Selecting items and constructing a test in this way, however, has the undesirable effect of automatically limiting the number of students who can achieve mastery on the test. Statistics used in constructing and evaluating tests rely upon the fact that the item and test scores vary widely and are normally distributed. These techniques of test construction, therefore, cannot be applied to criterion-referenced tests.

There are three steps in the development of criterion-referenced tests (such as flying exercises):

1. *Specify instructional objectives prior to instruction.* As has been previously stressed, effective instruction in content means that the objectives for each learning episode need to be determined and stated clearly beforehand. Each objective should contain an operational verb: one that describes what the student must do to demonstrate that he or she has learned. The essence of criterion-referenced evaluation is in demonstrating that the student can, or cannot, meet the objectives that have been prescribed.

2. *Decide on criteria that define adequate knowledge and/or performance.* Deciding whether or not a student has met the objectives is essentially a yes/no matter. The logical criterion is perfection—but humans always make errors, and so a more realistic question would ask *how much* error is permitted. It has become recognized that mastery in basic skills falls short of perfection by 10 to 20 percent (it is usually acceptable if the student is correct 80 to 90 percent of the time). This figure is arbitrary, however, and to a large extent depends upon the instructor's judgment or legislative requirements.

It is rarely a simple case of "If 80 percent or more correct—*Pass;* if 79 percent or less correct—*Fail,*" but rather, "If 79 percent or less—*Why?*" The answer then becomes an instructional objective in itself. We would drill the student in the areas of weaknesses. When that had been successfully accomplished, the student would be well into the 90 percent mastery zone (if that was the only difficulty).

Sensible criterion-referenced testing, therefore, does not only give a final pass-fail statement or a formula leading to that. It provides a list of diagnostic checkpoints that will give information as to why the goal wasn't reached satisfactorily. Sensible instruction demands that students should know what kinds of errors they are making, be instructed in their rectification, and permitted another go. The instructor using criterion-referenced evaluation has therefore to be very analytical about what is involved in teaching a certain exercise, in selecting aspects that will test that exercise, and structuring those aspects in a test.

It is useful to compare the construction of norm- and criterion-referenced tests. In criterion-referenced testing, one is interested in sampling the capabilities of students as they work progressively towards the criterion task: items are selected in terms of their logical relationship to the final instruction objective. A "good" item is one that is most related to instruction (before instruction, only 10 percent of students could pass the item, but after instruction the figure is 90 percent). A poor item would be one that might be passed by 50 percent before instruction and 60 percent after.

In norm-referenced evaluation, on the other hand, items are selected in terms of both how well they correlate with each other and thus measure the same thing, and how well they distinguish between good and poor students. As we have seen, a large initial number of items is selected, administered, analyzed, and discarded. The resulting test is internally consistent and distinguishes between good and poor students.

3. *Devising testing situations.* The final step involves committing the objectives to some form of test format. A test format may involve pencil-and-paper items (as in the usual objective test) or, because instructional objectives should be behavioral where possible, use of a practical context in which student behaviors are evaluated in flying exercise.

In summary, there are certain difficulties in obtaining reliable essay ratings, but it is possible to do so if one can afford to devote a rather large amount of time to marking and checking. Because of the peculiar benefits of the essay—not least that it allows the student to indulge in continuous writing—this time is usually worth investing.

A Comparison of Evaluation Methods

Now for an overall view of the media we have been discussing, with a quick look at the advantages and disadvantages of each. Comparisons are made in terms of both general properties of all tests, and some educational or practical points. These are set out in Table 6.1.

PROPERTIES OF TESTS

There are four general properties of tests to be considered: validity, reliability, distortion, and suitability.

Validity. The validity of the essay depends on whether examiners set unambiguous questions, and whether they mark in terms of the qualities they think they are marking. The short answer is probably a valid medium as far as it goes, but it is more restricted in scope than either the essay or the objective. The practical has excellent face validity: it makes clear sense to both student and evaluator, although there may be difficulties in reproducing the terminal situation exactly. The validity of the interview depends much upon the participants.

Reliability. The oral is the least reliable, followed by the essay, then the short answer, with the objective being most reliable. The practical can be very reliable if the behaviors are scheduled in advance on a checklist: if, however, general ratings of performance constitute the method of scoring, then the practical would be on a par with the essay.

Distortion. The question of distortion is complicated by the fact that so little research has been done on it, despite the fact that it is of great practical concern to both teachers and students. Distortion is not always bad. If the evaluator is aware of the kinds of distortion most likely to be influential in a particular format, then the evaluator can capitalize on this. This applies particularly to essay marking. If the evaluator knows the kinds of things to reward in students, which are independent of strict content-related material (for example, ability to write fluently and wittily), this should be made clear to students. It is also important that markers become aware of undesirable things they mark up (reproduction of teacher's words), and desirable things they mark down (originality). Distortion in the practical and the oral have not been researched, but most teachers have a

TABLE 6.1. Comparison of Tests

	Essay	Objective	Short answer	Practical	Oral
General properties					
Validity	Variable: depends on marker	Good if test well constructed (e.g., using a taxonomy and pretest)	Restricted but good as far as it goes	Excellent face validity	Can be good: depends heavily on participants
Reliability	Not so good, but can be increased with care	Excellent if properly constructed	Good, especially with model answers	Poor, if rating: good, if criterion checklist	Usually very poor
Distortion	Depends on marker, but can be employed usefully	Distortion present: favor convergent and associated	Not known	Probably present and varies according to situation	Considerable: depends on participants
Potential for norm- or criterion-referencing	Either; markers tend to mark normatively despite intentions. Ideally would include references to student's criteria	Either, but present test technology geared to norm-referenced	Either	Either; best suited to criterion-referenced	Very complex: usually turns out norm-referenced
Educational and practical points					
Difficulty in preparing	Little	Considerable; its greatest disadvantage	Little trouble	Should be (e.g., in preparing criterion checklist)	Depends on context
Coverage of course	Rather restricted	Excellent; can sample the whole of the course	Better than straight essay	As desired	As much as desired
Freedom for student to show abilities	Potentially free: depends on marker	Very limited	Compromise between essay and objective	Depends on context	Potentially free: depends on participants
Ease of scoring	Difficult: greatest disadvantage	Easy: greatest advantage	Good	Complicated, or should be, to make best use of situation	Difficult; beyond overall impressions

fairly clear idea of what they think are distorting factors (e.g., introversion-extraversion and anxiety in oral/social media).

Suitability of norm- or criterion-referencing. All media can be either norm- or criterion-referenced, but most turn out in practice to involve the former. In the case of objective tests, this is because the technology for norm-referencing has been around for so much longer. In the case of other media, all of which involve the judgment of the marker, there is a strong tendency to make comparisons between individuals if the criteria are not absolutely clear. The marker may start with every intention of rating the candidate in terms of particular criteria, but when pinned down to a decision, prevaricates: "Well, if I pass her, then I've got to pass Tom whom I already have decided should fail."

Everyone who has ever attempted evaluation knows how often it comes to this in the end, which is one reason why good criterion-referenced checklists are so useful. Each specific piece of behavior can be rated unequivocally and the outcome should be clear—and can be made clear to the candidate. In a global rating, the internal means of arriving at a decision so often involves purely personal like-dislike factors. Students are, of course, fully aware of this, and when they fail under a global-rating type of evaluation, they tend to feel that they have also been failed as a person—which may well be the case. Wherever personal judgment is involved, then, the evaluator should be extra careful about making clear what the student is really being rated on.

Essays and interviews also allow instructors to assess in terms of the student's criteria, if that is appropriate. Often students will interpret the question in their own ways: the marker may disagree, and in that case, must decide the extent to which the writer's view is justifiable.

PRACTICAL ASPECTS

There are several practical aspects of testing media, four of which (noted in Table 6.1) are dealt with here.

Trouble in preparation. Essays and short answers are the easiest of all tests to prepare, objective tests and practicals the most difficult. Oral interviews may be extremely easy or require a lot of preparatory work, depending on the context.

Coverage. One of the major advantages of objective tests is their enormous potential for coverage: items may be selected so that every important aspect of the course is sampled. At the other extreme is essay writing, which takes a considerable amount of time relative to selecting a multiple-choice item. While this means that a particular issue can be addressed in depth, it is difficult in the time available for examinations, to test coverage. Unless there is considerable choice offered in an essay exam, it becomes something of a lottery. The shorter time required on short-answer formats means that coverage here can be better than in the essay proper, but not as good as the objective test. Coverage in the practical should be close to total. If it is not, one of its major advantages (resemblance to a real situation) is lost. The interviewer can ask questions in depth or range widely (or both if time permits) providing a degree of flexibility, which is a great advantage.

Freedom for student to demonstrate relevant knowledge and skills. The objective format is clearly very limited in this respect. The student can either check the correct alternative or not. Rarely is the student permitted to explain reasons, offer alternatives to those presented, or justify the way questions have been interpreted. Indeed if this is permitted, one of the main advantages of the objective test (ease of scoring) is lost.

The essay, on the other hand, leaves a lot more up to the student. Students can, first of all, state and justify what it is they intend to do, and then do it. Freedom is given to orient the question so that it best reflects students' own thinking on the issue. Rigid marking of essays can of course make the medium just as restricted as an objective test. The short answer is more restricted than the essay, but less so than the objective test. In a well-conducted practical, the student ought to be required to show all relevant abilities (this is, of course, the purpose of a practical). On the other hand, there may be little freedom for the student if a criterion checklist is being used. If a student teacher has a particularly individual style of teaching that does not match the categories in the list, then no matter how effective that style could be, it will not be evaluated as effective. However, it would make little sense if the evaluator stuck rigidly to the checklist if all other indications were that the student was doing extremely well.

The interview is as free as the interviewer will allow it to be. The student can be left with no choice but to answer the questions and follow the leads given, or the student can be given complete responsibility for revealing himself or herself.

Ease of scoring. This factor probably accounts for the popularity of the objective test. Once the test has been set, the scoring is extremely quick and easy. This is in direct contrast to the essay, which is easy to set but difficult to score reliably. The short answer is a great improvement on the essay in this respect, although ease of scoring is offset by restricted freedom for the student. The practical is easy to score once the checklist has been well established, but it does mean that the evaluator has to concentrate long and hard while observing the student. Scoring an interview may be difficult, for if preparation is inadequate the interview often results in a general gut-level rating: the interview builds up a general feeling that the interviewee is excellent, good, passable, or not up to standard (each interviewer has a personal-peculiar calculus through which this decision can be reached).

These are some of the more important characteristics of the main forms of testing. The next important step in the evaluation procedure is to move from the test itself to the final distribution of grades.

GRADING

Implementing the results of evaluation involves that aspect of summative evaluation that causes most worries: grading. By grading, most people mean the award of labels (Distinction, Credit, Pass, Fail, or some such designation). There are, however, several alternatives to grading.

Written Evaluations

Written evaluations are qualitative statements made about each student for each subject: a criterion checklist, simple statements about each student's strengths and weaknesses, and so on. Such statements, if they are to be meaningful, need to be criterion-referenced, which also helps to avoid the subjectivity and unreliability of overall ratings. Such evaluations, however, take a long time to prepare and a lot of space to report. Their main value is for the student, the instructors, employers, or parents. They should be provided, regardless of what other forms of grading are used.

Self-evaluation

Self-evaluation is not the same as self-grading. In the latter, the student actually sets the grade, which may or may not then be averaged with the teacher's grade. In self-evaluation, students can evaluate their own work qualitatively (either in writing or in conference) in line with the students' own and the teachers' criteria. The teacher then incorporates the student's evaluation with the teacher's own written evaluations. Self-evaluation is in itself an important learning experience. The disadvantages are that pressure for high grades strains the student's honesty, and that in some subjects the student may not have the content expertise to make appropriate judgments. Once self-evaluation is put into context, however, it seems that it might be a component of most evaluational schemes.

The Contract System

Contracts between student and instructor can be devised either on an individual basis, whereby each student negotiates a contract, or on a class basis, where the terms of the contract apply to everyone in the class. In either case, it is agreed between students and teacher that so much work is worth an "A" (or whatever labels are used), so much a "B," and so on. Where a student fails to meet the terms of the contract, a new one can be negotiated or the outcome left to the teacher's judgment. A contract could consist of something like the following (in this case the teacher likes good attendance):

F: Failing to hand in any assignments, and failing to attend more than two-thirds of the classes.
D: Attendance satisfactory but assigned work unsatisfactory.
C: Attendance satisfactory; at least one satisfactory assignment.
B: Attendance satisfactory; at least two satisfactory assignments.
A: Attendance satisfactory; assignments unusually good or extra work undertaken.

Very similar to the contract is the point-accumulation system, according to which points are awarded for amount and quality of work. An A may be worth 10–12 points; B, 7–9; C, 4–6; D, 2–3; and F, zero or 1 point. There is no need for a student actually to sign a contract. The student knows the rules (and may well have a hand in deciding them) and keeps working until sufficient credit is gained for a grade that satisfies the student. (Under this kind of system, a student should be permitted to resubmit a poor paper.) Points can be awarded for quantity or for quality, but preferably for both (zero for unsatisfactory, 1 for pass, and 2 for excellent). This tends to avoid one disadvantage of contracting, that it can reinforce sheer quantity of output. It does mean, however, that the instructor's work load is much increased: a determined student can easily wear a teacher down with requests for resubmissions or detailed evaluations.

Since grading here is criterion-referenced, students understand where they stand, and that their grade is quite independent of anyone else's.

Mastery Learning

Mastery learning is not so much a grading procedure as a total instructional approach. Mastery learning can be used in connection with a letter-grade system by either defining mastery in terms of "0.0 percent error = A, 1–5 percent error = B," and so on, or by using one criterion for a variety of tasks (giving A if all tasks are complete, B if most are, etc.).

Pass/Fail

Pass/Fail is usually part-and-parcel of a criterion-referenced approach (student does or does not meet the specified standards), but it can stand as a grading method in its own right. Pass/Fail is where the only statement on a report or transcript is whether the student passed or failed the course. Its application in nonmastery courses is simply that the higher grades are all compressed into one overall "P." Variations of the method retain an "Excellent" category in addition to Pass and Fail.

Pass/Fail removes much of the pressure to compete for higher grades. However, some of the more ambitious students may not feel it worth their while to try as hard as they might.

Credit/No Credit

A version of Pass/Fail is the Credit/No Credit system, which is preferable in many ways. The student receives credit for passing a course, but there is no permanent record of failing. The point is that these attempts are nowhere recorded officially, which removes some of the pressure on the borderline student.

Summary and Conclusions

Several forms of testing are in common use, and each can be assessed in terms of its validity, reliability, and distortion. The main media are:

1. *The essay.* This is easy to set and has a good reputation for validity but a bad one for reliability. The main problem is that the marker's biases can distort the marks in many ways, thus distortion varies according to who is marking the test. However, both the quantitative and qualitative reliability can be improved with the use of component tables drawn up prior to marking.

2. *The objective test* (multiple choice). Difficult to set, if done properly, but very easy to score. Good reputation for reliability, but validity is often restricted. Some distortion is present in that it appears to favor convergent rather than divergent abilities.

Discussion Questions

1. From a flight training program with which you are associated, provide an example of a summative and a formative evaluation in both theory and practical components.
2. Name one flying exercise, then provide a criterion-referenced test that could be used to evaluate the learner's performance.
3. Which do you regard as the greater problem in flight instruction: the validity or the reliability of tests? Refer either to ground school or flight testing. Justify your choice.
4. When flying standards are being assessed, what are the main sources of measurement error?
5. Select a suitable topic from a training syllabus, then devise a 10-item objective test. Items should be of the true/false, matching, fill-in-the-blank, and multiple choice varieties.

6. You are to evaluate a junior instructor's preflight brief. What criteria will you use? How will you go about it? Justify your answers.
7. You are to evaluate a period of airborne instruction. Prepare a checklist that you will use as a basis for your assessment of the instructor. Justify your checklist construction.

Further Reading

Gagne, R. M. "Military Training and Principles of Learning." *American Psychologist* 17(1962): 83–91.

Roscoe, S. N., and J. M. Childs. "Reliable, Objective Flight Checks." In *Aviation Psychology* by S. N. Roscoe. Ames: Iowa State University Press, 1980. Roscoe and Childs examine a four-stage approach to flight checks: (1) establish explicit indices of desired performance for each training task, (2) observe and record a small number of critical variables at specified times or positions (such as 45 seconds into a standard turn to check if the aircraft is passing through 135 degrees), (3) make explicit the acceptable limits of performance for each task, (4) develop and exhibit criterion-referenced learning curves to aid assessment.

Thorndike, R. L., ed. *Educational Measurement.* Washington: American Council on Education, 1971. A collection of papers written by experts on different aspects of evaluation.

7 Aspects of Training and Instruction

THIS CHAPTER consists of four aspects of instruction in which most of the content of the preceding chapters can be applied. Behavioral objectives have been described previously as advance organizers and as a basis for the evaluation of flight instruction. In this chapter, examples are given of their application to the planning of flight instruction. The advantages and disadvantages of their use are considered.

Reflective instruction is a means of linking an instructor's craft with teaching skills. Reflective instruction has proved to be successful in sensitizing instructors to the need for professional skills in order to facilitate student learning. To enable readers to see for themselves, a sample lesson is included.

Then comes an exposition of pilot judgment training, which is currently undergoing large-scale trials internationally. There are expectations that judgment training can be integrated with the usual pilot training so as to produce significantly improved judgments related to safety in aircraft operation.

Finally, there is a brief analysis of approaches to training in cockpit resource management. Basic approaches include the use of the managerial grid and situational leadership.

The following questions are answered in this chapter: What are the advantages and disadvantages of using behavioral objectives? Where can an instructor find a summary of teaching skills that apply to flight instruction? What is airmanship? How can pilots be trained in making safe judgments? What is meant by the "maturity" of flight crew involved in decision making?

After reading this chapter, you should be able to: (1) write a behavioral objective for either a ground-school subject or a flying exercise; (2) conduct a reflective teaching lesson; (3) define pilot judgment; (4) provide principles of flight deck management; (5) determine when subordinates should be involved in decisions; and (6) decide on an appropriate leadership style to suit the situation and the subordinates involved.

Part A. Behavioral Objectives

Chapter 1 presented a basic instructional plan. Its three elements consisted of content, rationale, and presentation. All three are related to the training objectives that enable pilots to operate aircraft. S. N. Roscoe argues that effective instruction should be based on a categorical analysis of the pilot behaviors being sought. These include procedural activities (such as managing communication, navigation, fuel, and power), perceptual motor activities (such as geographic orientation and communication), and decisional activities (such as navigation planning or crew functions).

How can such behaviors be expressed in an accurate way to guide instructional planning, presentation, and evaluation? A general strategy is shown in Fig. 7.1, which establishes a sequence that begins when an instructional objective is defined. A pretest establishes whether or not the learner can achieve the objective. If so, the instructor redefines the objective to suit the following instructional exercise; if not, instruction is given and learning is subsequently tested by applying specified criteria.

The critical component in the sequence of evaluation and instruction is the *objective.* It is derived from more general goals provided by either a syllabus, regulatory requirement, or training policy. The objective is a statement of the performance expected of the learner at the end of the instructional period. By specifying the objective in a systematic and consistent way, the training pilot can simplify the task of evaluation.

WHY BEHAVIORAL OBJECTIVES?

Very general statements of aims can be of little assistance to the flight instructor or the trainee pilot in the instructional period. For example, how do you determine competence in a student pilot to the extent that you could regard the student a "good" pilot? Each flight instructor could no doubt provide a list of criteria. But to what extent would the criteria be common and to what extent could one instructor use the criteria of another?

A less-likely statement of aims may exemplify the point more effectively. Let's set out to teach someone "all about flying." How could we determine if this aim had been accomplished? Writing tests, undertaking a rigorous flying test, answering questions, demonstrating particular techniques, simulating responses to emergencies, responding in a discussion—all are possible ways of evaluating the ultimate pilot. But do any or all of these tests indicate that the pilot knows "all about flying"? The solution to this problem lies not in the test, but in the way the objective is expressed.

The proper definition of a learning objective can help the training pilot, the trainee, and the teaching-learning process. Definition of objec-

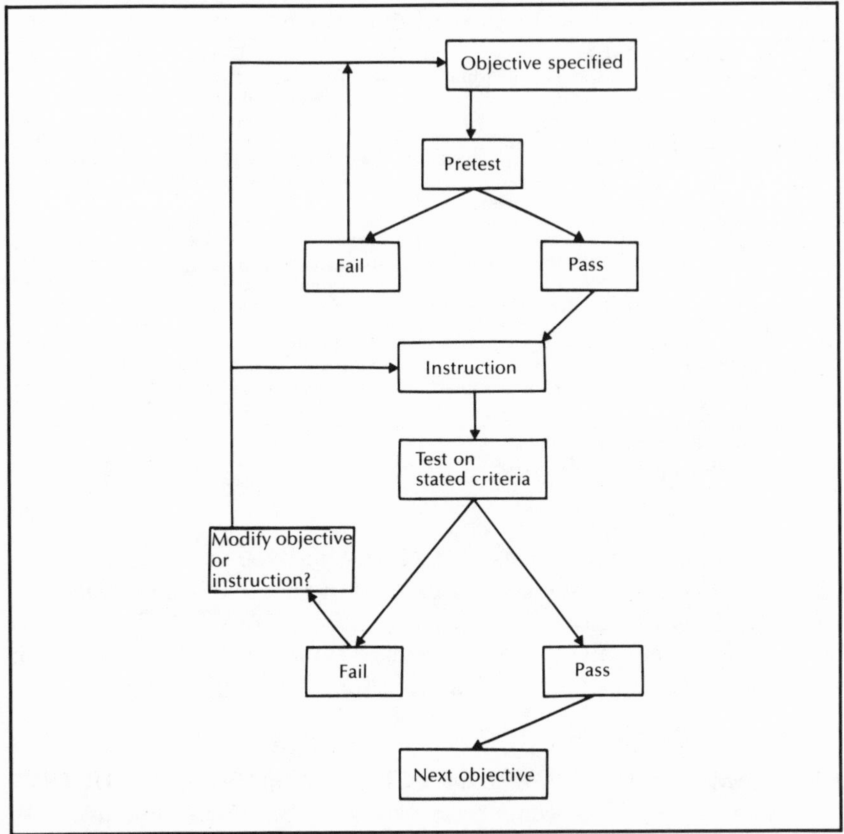

FIG. 7.1. **Using Objectives in Flight Instruction**

tives in behavioral terms is by no means a newfangled hobbyhorse of academics: it can be traced back to 1924 publications by Franklin Babbit. Since that time there has been considerable controversy over the use of behavioral objectives. However, we should examine the arguments for and against their use. Table 7.1 provides a balanced view.

There can be no doubt that the instructor or check pilot who seeks to utilize behavioral objectives will be, at least initially, committed to more time and effort in instructional planning. Is it worth it? Many benefits, apart from those listed in Table 7.1, have been argued. Communication is a major strength, for behavioral objectives can inform potential clients, departmental officers, or the general public of a flying school's intentions; teaching goals can be transmitted to other instructors; and learners can be informed of their instructor's expectations. Instructors themselves are as-

TABLE 7.1. **For and Against Behavioral Objectives**

ADVANTAGES	DISADVANTAGES
1. They increase teacher awareness of what the student should be working for and lead to more optimal planning; a wider range of objectives is included.	1. They increase teacher control at the expense of students' own objectives.
2. They provide a basis for assessing continuous progress (allowing students to proceed at their own rate) as specific skills are focused upon.	2. They unduly emphasize things that can be measured; only low-level objectives are likely to be measured.
3. Students have a better blueprint for guiding their learning activity; hence, they learn more.	3. Once established (teacher invests the time to write objectives and set up a measurement system), the system perpetuates itself; objectives remain the same; spontaneity is reduced.
4. After reaching a criterion, students have more time to work on their own learning objectives.	4. A great deal of instructor time is used in writing behavioral objectives; such time could be utilized better in other ways (such as one-to-one conferences).

Source: T. L. Good, B. J. Biddle, and J. E. Brophy, *Teachers Make a Difference* (New York: Holt, Rinehart and Winston, 1975), 147.

sisted in their selection of aids or materials, test construction, and wider planning of not only exercises, but units or courses within a syllabus.

Behavioral objectives have also been seen as a means of correlating activities and materials to specific learning situations, and as a means of assessing the success of a course, unit, lesson, flight, or exercise.

CRITICISMS OF BEHAVIORAL OBJECTIVES IN FLIGHT INSTRUCTION

Following are some common criticisms of behavioral objectives. We think that as far as flight instruction is concerned, these objectives can be met, as indicated below:

1. *Important outcomes will be underemphasized because lesser outcomes are easier to present in behavioral terms.*

It is patently easier to state explicit, behavioral objectives for stall recovery than for pilot judgment or airmanship. However, the fact that these objectives are explicit makes it easier for instructors, their superiors, and colleagues to scrutinize them and eliminate those that are inappropriate.

2. *Specifying objectives prevents instructors from taking advantage of opportunities unexpectedly occurring in flight exercises.*

Specifying ends does not imply specified means. It is a reasonable constraint that instructors have to justify digressions in terms of their contribution to an instructional objective. Gifted instructors typically capitalize on the unexpected, and such instructors are unlikely to be adversely affected by the use of behavioral objectives.

3. *Pilot attitudes and values are just as important as observable changes in behavior.*

Agreed, there are some aspects of instruction that are very difficult to express in specific behavioral terms. The ultimate criterion of effective training is the pilot's performance: isn't it reasonable that we establish this as the basis for evaluation?

4. *Mechanically measuring behavior is a dehumanizing approach to instruction.*

Like Olympic judges of diving or gymnastics, experienced instructors are capable of fairly reliable judgments based on qualitative, rather than quantitative, grounds. Independent evaluations can be quite similar.

5. *It is undemocratic to plan a learner's behavior after instruction.*

In the 1960s programmed instruction was criticized in this way. But instruction is inherently undemocratic as it is a set of deliberately structured experiences. With such a high investment in equipment and human expertise, it is difficult to see an alternative.

6. *Let's be realistic. Flight instruction isn't like that.*

There is a difference between describing the status quo, and advocating it. **What** often differs from **what should be.** Do you regard the current nature of flight instruction to be optimal?

7. *Certain instructional areas, such as decision making, are problematic when one attempts to identify measurable student behaviors.*

Agreed, but let's make a start (if only tentatively) on clearly defining criteria.

8. *General statements of aims appear more worthwhile than very precise objectives.*

Behavioral objectives can be a threat, forcing the instructor to defend teaching content. Shouldn't flight instruction be subject to scrutiny and justifiable criticism?

9. *Instructors may find themselves accountable in terms of their ability to produce these identifiable results in trainees.*

Shouldn't they be accountable? The advantage, however, is that the use of behavioral objectives establishes a common ground for evaluation. A senior officer cannot use the way he or she used to instruct as a criterion. Evidence of student attainment is available. If it isn't, it should be.

10. *Generating behavioral objectives is too time consuming.*

It could be. The lists of verbs in this chapter should help. Flying schools could quickly build up a data bank of objectives available to practicing instructors and those undertaking refresher courses.

11. *Prespecified goals may diminish the importance of the unforeseen, but critical, results of instruction.*

Really dramatic and unanticipated outcomes could hardly be overlooked by instructors. If there is reason to believe that there will be a certain outcome, it should be part of the planning and evaluation. See, for example, the use made of objectives in evaluating turn maneuvers in Fig. 7.2.

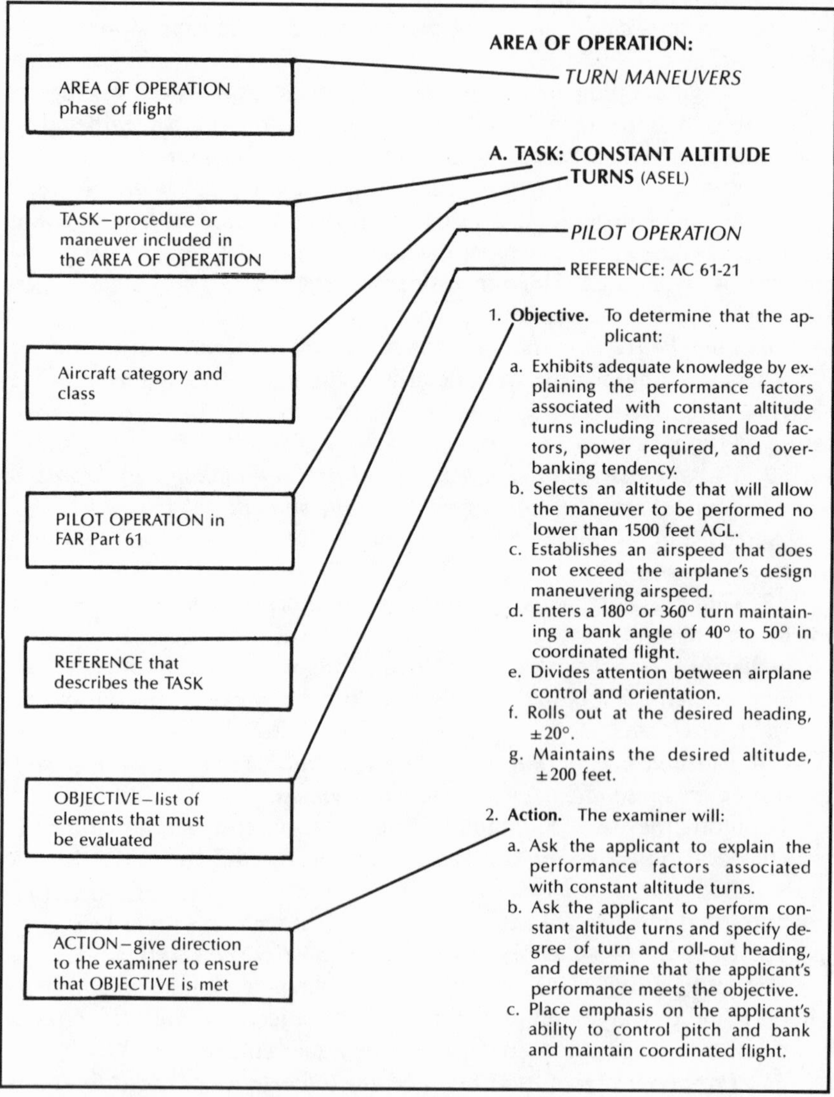

FIG. 7.2. **An Approach to Examination Based upon Behavioral Objectives**

USING OBJECTIVES IN FLIGHT INSTRUCTION, CHECK AND TRAINING

Naturally, a flight selector's choice of lesson plan does not only depend on objectives. Other relevant considerations include resources (both human and material), facilities, the individual, and the program. After taking these factors into account, the instructor's *goal* (a broad, general statement of learning outcome, such as "to produce a competent, efficient, safe pilot for the operation of a PA28 in appropriate conditions") leads to the *objectives* (which are more specific statements of learning outcomes, always expressed in terms of the learner's performance). This process is traced in Fig. 7.1, which shows how the cycle of instructional planning goes from one objective to the next.

Objectives are specific, rather than general; expressed from the student's point of view; expressed in performance terms; and indicators of the means by which they will be evaluated through performance criteria.

Objectives can be expressed as terminal behavioral objectives that specify desired functional performances *after* instruction. Again, these will be both specific and expressed from the learner's viewpoint. Additionally they will be expressed in terms of the intended performance of the student. It may also contain references to special conditions or criteria related to the performance. For example, a flight instructional objective could be "Given the use of ground references to control path, the trainee will demonstrate compensation for wind drift following a specified pattern of attitude, altitude, and course."

In order to express objectives from the student's viewpoint it is necessary to avoid phraseology such as:

To train the student . . .
To aid the student . . .
To develop in the student . . .
To help the student to become . . .
To bring about in the student . . . and so on.

Consider the examples of student-oriented behavioral verbs (Insert 7.1) suitable for use in the specification of objectives.

Insert 7.2 shows the progression of an abstract or general goal, through a subgoal (which is still quite general) to a second subgoal (now expressed in observable but problematic terms) to a third subgoal, which is expressed as a behavioral objective. It meets the requirements for a behavioral objective because it indicates the condition in which the learner has to produce the performance, the minimum standard of performance that would be acceptable, and the manner in which the accomplishment is to be evaluated.

INSERT 7.1. **Behavioral Verbs for Use in Flight Instruction**

Mathematical Behaviors
Add Bisect Calculate Check Compute Count Derive Divide Estimate Extract Graph Group Measure Multiply Number Plot Prove Reduce Solve Square Subtract Tabulate Tally Verify

Instrumental/Scientific Behaviors
Calibrate Connect Convert Decrease Demonstrate Increase Insert Length Limit Manipulate Operate Plant Prepare Remove Replace Report Reset Set Specify Time Transfer Weigh

Creative Behaviors
Alter Ask Change Design Generalize Modify Paraphrase Predict Question Rearrange Recombine Reconstruct Regroup Rename Reorder Reorganize Rephrase Restate Restructure Retell Revise Rewrite Simplify Synthesize Systematize Vary

Complex, Logical Judgmental Behaviors
Analyze Appraise Combine Compare Conclude Contrast Criticize Decide Deduce Defend Evaluate Explain Formulate Generate Induce Infer Plan Structure Substitute

General Discriminative Behaviors
Choose Collect Define Describe Detect Differentiate Discriminate Distinguish Identify Indicate Isolate List Match Omit Order Pick Place Point Select Separate

Language Behaviors
Abbreviate Alphabetize Articulate Call Outline Print Pronounce Read Say Sign State Summarize Tell Translate Verbalize Write

"Study" Behaviors
Arrange Categorize Chart Circle Cite Classify Compile Copy Diagram Father Find Follow Itemize Label Locate Look Map Mark Name Note Organize Quote Record Reproduce Search Sort Underline

Flight Verbs
Accelerate Acknowledge Aim Ascend Bank Begin Bring Climb Comment Compensate Complete Conserve Consider Control Correct Criticize Decelerate Descend Designate Determine Develop Discover Distribute Drop Emphasize Evaluate Explain Extend Feel Finish Fit Fix Follow Guide Hand Hold Imitate Increase Initiate Lengthen Lower Maneuver Miss Observe Open Plan Position Practice Produce Provide Raise Recognize Recover Reduce Regain Relate Repeat Retain Return Slide Slip Shorten Slow Stall Supply Support Start Switch Take Touch Try Turn Use Watch Work

In specifying objectives, evaluation is aided by the inclusion of performance criteria, which refer to time permitted, standards of accuracy, aids permitted, or other restrictions. For example:

Time: per lesson, per hour, time in the air, within thirty seconds

INSERT 7.2. Steps from General Goal to Behavioral Objective

GENERAL GOAL	SUBGOAL (1)	BEHAVIORAL SUBGOAL (2)	OBJECTIVE SUBGOAL (3)
To acquire a basic understanding of aircraft controls.	To understand the ways in which an aircraft is maintained in straight and level flight.	To regain straight and level flight from a turn, climb, or descent.	After an introduction to the effects of controls, the student will be able to return the aircraft to straight and level flight from slight turns, dives, and climbs.
COMMENT:			
Possibly observable, perhaps, to another flight instructor or evaluator.	Instructor knows requirements and can observe students: but how is it assessed?	Observable but difficult to evaluate standard of performance, on comparability.	Gives condition of performance, minimum standard of achievement, and suggests how accomplishment can be measured.

Standards: with 100 percent accuracy, from memory, with the use of manuals, in no more than two pages, seven out of ten
Aids: provided with a model aircraft, map, overhead transparency, video, 35mm slide, carburetor, compass
Restrictions: from memory, without use of instruments, by instruments only

A behavioral objective for map reading could be "Students are to develop skill in identifying map references given latitude and longitude readings. For each pair of coordinates the student will write a name, making use of available maps."

Thus a terminal behavioral objective can be judged according to its (1) specificity (note the way in which the objectives in Insert 7.2 become more specific); (2) expression from the student's viewpoint; (3) content of performance statement and, perhaps, reference to special conditions or standards.

TAXONOMIES OF OBJECTIVES

A taxonomy is a scheme for classifying a group of entities. For example, a taxonomy of aircraft might classify according to purpose, load, or power. Because educational objectives are designed for education in all its forms, taxonomies of educational objectives need to be quite comprehensive (see Further Reading). Taxonomies of learning objectives have been available for some time and provide a useful reference, especially for the design of instructional programs. They assist the designer to rank objectives in order of difficulty for student attainment.

SUMMARY

Behavioral objectives are versatile and potent devices for planning, presenting, and evaluating instruction. Despite their limitations in areas such as airmanship or attitudinal aspects, they are especially suited to flight training where they provide a precise focus and remove any conflicts in expectations. More importantly, behavioral objectives provide an integrated means of evaluating instruction and learning.

Discussion Questions (Part A)

1. Determine if the following objectives may be classified as behavioral objectives. Rewrite any objective that is not a behavioral objective.

   (a) Given an out-of-balance aircraft in flight the student can restore balance and trim for straight and level flight with single use of each trim control.
   (b) The student will accurately identify the principle axes of an aircraft on a model supplied.
   (c) On the chalkboard, the student will accurately draw a parallelogram of forces analyzing thrust and drag components in an aircraft.
   (d) The behavioral objective is to instruct the student so that, given a model, the student can accurately identify the principal axes.
   (e) To enable the student to draw, with 100 percent accuracy, a parallelogram of forces of thrust and drag components.

2. For each of the following exercises, develop a subobjective and express it as a behavioral objective suitable for use in flight instruction.

   1. Aircraft Familiarization
   2. Preflight Operations
   3. Familiarization Flight
   4. Use of Radio for Communications
   5. Straight-and-Level Flight, Turns, Climbs, and Descents
   6. Slow Flight and Stalls
   7. Coordination Exercises
   8. Ground Reference Maneuvers, Traffic Patterns
   9. Takeoff and Departure Procedures

10. Approach and Landing Procedures
11. Stalls from Critical Flight Situations
12. Steep Turns
13. Cross-wind Takeoffs and Landings, Slips
14. Short- and Soft-Field Takeoffs and Landings, Maximum Climbs
15. Power Approaches to Full Stall Landings, and Wheel Landings
16. Turns to Headings and Recovery from Unusual Attitudes (using instrument references only)
17. Emergencies
18. The Solo Flight
19. Practice Area Familiarization
20. Cross-country Flight Planning
21. Pilotage, Map Reading
22. Dead Reckoning, Use of the Compass
23. Use of Radio Aids for VFR Navigation
24. Obtaining Emergency Assistance by Radio
25. Unfamiliar Airport Procedures
26. Plotting Alternate Courses in Flight

Part B. A Test of Instruction

Paradoxically, one problem in flight training may be the skill and experience of the instructor or training officer. Sometimes those with great expertise do not know how to relate that expertise to others. The fact that one can achieve high standards of operations oneself does not automatically confer an equivalent ability to teach others to achieve the same level.

This view is reinforced when highly skilled pilots are asked to undertake some simple teaching tasks that are completely unrelated to flying. Instructors quickly become aware of the need for teaching skills when their lesson is on origami or a fantasy topic such as a devised language. A valuable method of taking instructors out of their area of expertise and forcing them to concentrate on the processes of teaching and learning is termed *reflective teaching,* the name being derived from the subsequent consideration and discussion of the teaching by a small group of the instructor's peers. Use of this teacher assistance technique can help an instructor become more effective.

REFLECTIVE TEACHING

Consider a typical instructor refresher course attended by, say, 15 instructors. Prior notice would have been given to 3 of the instructors who would have been issued a set of notes such as that which follows (Insert 7.3). Those instructors would have been asked to prepare the short lesson to achieve both the stated objective and learner satisfaction.

INSERT 7.3. Reflective Teaching Lesson #1
Practicing Describing Behavior Using the Block Diagram

Description of the reflective teaching task

You are to describe a visual object so well that your students can draw the object without seeing it. Plan to teach this lesson in a way that will result in both student learning and satisfaction.

Introduction to the lesson

Instructors need to describe things effectively—that is, use words accurately—to ensure that they are conveying precise meaning to their students. Below is an objective that requires you to describe something to a small group. The task was selected because your success in accomplishing it probably will not be dependent on your knowledge, previous experience, or flying qualifications. Rather it will exercise your ability to describe carefully.

Objective

Your goal is to get as many of your learners as possible to produce an exact replica of the blocks as shown in the lesson diagram. You will have ten minutes in which to accomplish your goal.

Limitations

1. *You may only use words when you teach the lesson. You may not show the learners a picture of the blocks.*
2. *Seat your learners so that they are in a circle but facing outward so that they will not be influenced by each others' work.*
3. *You may not use any special devices such as rulers.*
4. *You are limited to ten minutes.*

Ending the lesson

1. Notify the chief flying instructor as soon as your learners have completed the task. The instructor will record the time.
2. Collect the block diagrams.
3. Obtain copies of the Learner Satisfaction Forms from the chief flying instructor.
4. Distribute the forms to the learners advising them that they will have two minutes to complete them.
5. While the learners complete the forms, check the block diagrams.
6. Fill in the Scoring Box, which follows.
7. Collect the Learner Satisfaction Forms.
8. Return the corrected block diagrams to the learners.

▶

▶

Scoring Box

Place an X in the box if the learner succeeds in producing an exact replica of the lesson diagram, a zero if not.

Learner ______________________________

1 ______________________________ ☐

2 ______________________________ ☐

3 ______________________________ ☐

4 ______________________________ ☐

5 ______________________________ ☐

Number of X's

Lesson Diagram

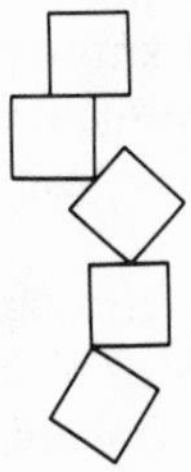

Learner Satisfaction Form

1. During the lesson how satisfied were you as a learner?

very satisfied	satisfied	unsatisfied	very unsatisfied

2. What could your teacher have done to increase your satisfaction?

The chief instructor would quickly allocate the remaining 12 to the 3 groups of learners who would move to different parts of the same room for their lesson. The time of 10 minutes would be strictly enforced, and would be followed by a short test (provided with the lesson, but retained by the chief instructor and distributed only after the lesson has been given). In the test, the learners would be evaluated on their attainment of the lesson objective and asked to indicate their satisfaction with the way the lesson was presented.

The instructor of each group collects and collates the test responses, which are returned to the learners. Ten minutes is then allocated to a discussion (reflection) of the lesson, and what instructor and learner perceived as educational insights from what occurred. Then the whole group reconvenes for further reflection, combining the group views in a plenary session of 10 or 15 minutes.

BENEFITS OF REFLECTIVE TEACHING

Such reflection frequently leads to a rediscovery of the benefits of working from the known to the unknown, from the simple to the complex, from the concrete to the abstract, from the particular to the general, from observations to reasoning, or from the whole to the parts and back to the whole. Basic sequences (such as preparation, explanation, demonstration, imitation, practice, and testing), which are crucial to flight instruction, come under scrutiny. The demanding time constraint forces the instructor to identify the essential core of knowledge that must be known by the learner, separating this from the valuable or interesting aspects that should be known and the more extraneous aspects that could be known if only time permitted.

Most importantly, the instructor experiences the learner's perspective again and commences an analysis of teaching from that viewpoint. As the developer of reflective teaching, Donald R. Cruickshank, put it: "Rather than behaving according to technique, impulse, tradition and authority, students of teaching would come to deliberate their teaching with open-mindedness, wholeheartedness and intellectual responsibility."

Reflective teaching is not merely an aspect of preservice preparation of flight instructors: it is a means of regularly reminding instructors of the need to maintain a thoughtful ongoing analysis of their day-to-day instruction.

TEACHING SKILLS

Reflective teaching will frequently bring about a desire to investigate sources of teaching skills that will assist an instructor to be both more effective and more efficient as a teacher. A highly suitable reference source is the set of books and videotapes entitled *Sydney Micro Skills* (see Further Reading). It provides the components of skills, such as introducing and concluding lessons, explaining, questioning, reinforcing, individualized instruction, and so on.

For the instructor who has experienced problems in teaching method, the analysis of skills is very helpful. For example, the *skill of explaining* is first analyzed in terms of its components:

1. *Clarity,* which is enhanced by asking single questions, being explicit, and avoiding vague expressions
2. *Using examples* to clarify, verify, or substantiate concepts by
   (a) starting with simple examples and progressing to the more complex
   (b) starting with examples relevant to students' experience
   (c) using positive and negative examples of a concept
   (d) relating the examples to the principles or ideas being taught by using specific linking words
   (e) checking to see if the lesson objectives have been achieved by seeking examples, illustrative of the main points, from students
3. *Emphasis* to reinforce major ideas, key words, or principles; major techniques are variation (voice, gesture, movement, use of media, and materials) and structuring (repeating main ideas in various forms and by verbal cues)
4. *Feedback* from the full range of students by a variety of techniques

Next the modes of usage to individuals, part, or all of a group are discussed. Finally, the principles of usage (adequate planning and suitable analysis of the topic) are provided.

For the flight instructor, there are practical pointers in the full range of teaching skills. Aspects that are covered include the use of the pause in questioning; distinguishing between broad and narrow focus in questions; redirecting questions to minimize the instructor talk; and prompting by the rephrased question, asking simpler questions, or reviewing the relevant information before repeating the question.

In summary, reflective instruction can often serve to make an experienced instructor aware of the need for a knowledge of and expertise in the use of teaching skills. With the realization of that need, the instructor can then go to sources such as *Sydney Micro Skills* to find a distillation of professional skills needed.

Discussion Questions (Part B)

After completing the reflective teaching lesson and collating the results, return the tests to the learners and ask the following questions. Record a summary of the responses.

1. As learners, how satisfied were you with the lesion (very satisfied, satisfied, unsatisfied, or very unsatisfied)?
2. The results of the test on attainment of the lesson objectives were. . . . . What, in your view, contributed to this result?
3. The lesson was based on description: telling about something using words. From a learner's viewpoint, what are some of the important principles about instruction involving description?
4. As an instructor playing the role of a learner, what did you learn most from the experience?
5. If you had the task of teaching this same lesson tomorrow, how would you change the content and presentation from that you just experienced? Why?

Part C. Pilot Judgment Training

The essence of "airmanship" is defined by Amir Mané of the University of Illinois this way:

> Airmanship means not asking the control tower for permission to do something when it is obvious that they would not grant the request (e.g., permission to enter the runway while there is another airplane in finals).
>
> Airmanship means to be able to drive your car in a very long street without stopping even once at the lights.
>
> If you come to the base through a point where the height of reporting is 6,000 feet and the approach control tells you to report at 7,000 feet, good airmanship means to start looking for the other airplane which is most probably ahead of you somewhere.
>
> When exercising right above your base (and according to the approach procedure, you would have to make a long detour), good airmanship is to get permission to exercise emergency landing and in this way get straight down to base.
>
> Airmanship is the ability and the tendency to define the relevant aspects of an aerial problem or opportunity.

Airmanship is a slippery concept; however, Mané has been able to grasp it long enough for us to identify some of its dynamics. Pilot decision making and judgment is a similar topic. Few would deny its importance, but few would agree on a definition. A basis for agreement however, would be the view that decision making and judgment are aspects of airmanship. Further agreement would be gained on the view that most general aviation accidents are the result of faulty pilot judgments, therefore questionable airmanship.

Richard Jensen in the essay "Pilot Judgement: Training and Evaluation" points out that judgment can refer to highly learned perceptual responses made in quick order, or to the mental activity in choosing from alternate courses of action as in "staying ahead of aircraft."

STUDIES OF PILOT JUDGMENT TRAINING

A series of studies have been made of pilot judgment training, commencing with the experiment conducted at Embry-Riddle Aeronautical University in 1981 for the U.S. Federal Aviation Administration. The following year Transport Canada conducted a study at four aero clubs. In both experiments pilot judgment behavior, as indicated on test flights, was found to be significantly superior to control groups of students who had not received the training.

The third study was undertaken at ten fixed operators in the United States (see Diehl 1985), and another has been completed in Australia by Telfer and Ashman (1986) showing significantly improved safe flying practices among pilots given judgment training. In the last study, the Transport Canada definition of pilot judgment was adopted for the trial study (see Lester, Diehl, and Buch 1985). Pilot judgment was defined as:

> The process of recognizing and analyzing all available information about oneself, the aircraft, and the flying environment, followed by a rational evaluation of alternatives to implement a timely decision which maximizes safety.
>
> Pilot judgment, therefore, involves one's attitude toward risk taking and one's ability to evaluate risks, and to make decisions based upon one's knowledge, skills, and experience. A judgment decision always involves a problem or choice, an unknown element (usually a time constraint), and stress.

Pilot judgment training is based on the theory and practice related to recognizing and overcoming potentially hazardous attitudes and thought patterns. The method of instruction incorporates manuals for the student and instructor, briefings, and in-flight exercises. The student is given scenarios involving poor judgments and assistance with ways of circumventing the poor judgment chain. An analysis of stress is provided, together with its relationship to pilot capability and performance. Symptoms of stress and methods of overcoming stress are described. In instructional exercises, pilot judgment training is also related to the preflight check and aircraft systems, official procedures and communications, cross-country flying and night flying.

Pilot judgment training occurs naturally within most flight instruction. There is value in structuring and formalizing its place as an integral part of flight training.

Flight instructors can teach students to assess and cope with risks associated with flying by following a five-step process. This procedure, disseminated by the AOPA Air Safety Foundation in 1986 (see Lawton, Clarke, and Benner), recognizes that pilots receive little training in identifying problems (such as those associated with the hazardous attitudes emphasized by pilot judgment training). The five-step process that is taught by instructors and practiced by pilots is:

1. Recognize changes that affect flight outcomes.
2. Assess the risk(s) involved due to changes.
3. Decide what action(s) would control the risks.
4. Take action according to the decision(s) reached.
5. Monitor the effectiveness of the actions.

Risks associated with the pilot, aircraft, environment, and time can be assessed by using a rating scale. Risks can be balanced on an imaginary balance on which the fulcrum is the beginning of the scale (i.e., zero). The outer end of the beam can be given the maximum score (say, four), and all risks can then be allocated a place on the scale (zero to four). The pilot has to keep the beam balanced by reacting to oppose the "weight" of risks.

Consider this example (condensed from the AOPA safety report):

A non-instrument-rated private pilot and family are returning from a weekend in New Orleans to their home in Houston, Texas, by a familiar route. The pilot calculates that a nonstop flight at 8500 feet will provide a 45-minute fuel reserve. The forecast is for a weak front near Houston with ceilings above 2000 feet and possible isolated afternoon thunderstorms. At the cruising altitude these storms should be visible and avoidable, decides the pilot.

Servicing delays departure by two hours until 5:00 p.m., but by 7:30 p.m. they are halfway home. Lightning is flashing in the distance, illuminating a large area of thunderstorms. It is dark, and the airplane unexpectedly flies through the edges of a few clouds forcing the pilot to fly by instruments to return to VFR conditions. Now it is time for assessment, for the judgment chain has begun.

On a scale of 0–4, each of the four major variables in the situation can be assessed for risk.

The first consideration is the aircraft. There are no problems with equipment, but cloud-dodging has exceeded planned fuel consumption. This problem will be exacerbated if a descent is made below the clouds where lower true airspeed and groundspeed will result. The aircraft risk is rated as 2.

Next is the environment. Worsening weather means IFR and that is, simply, illegal. The weather is a major factor as it is an unknown ahead and is adversely affecting the passengers. The environmental risk is rated as 4 because of the lightning, clouds, and family anxiety.

Then there is the pilot who is not instrument rated, and last had a flight with an instructor over a year ago. There is pressure on the pilot coming from the need to get home, to get the children back to school, and to meet business commitments. The pilot risk is rated as 3 as he is current, but not instrument rated.

And, finally, there is the subject of time. In association with fuel consumption, the risk factor increases as fuel time remaining decreases. This geometric progression shown by the risk when there are 2 hours remaining (say, a factor of 1) compared with the risk when there are only 20 minutes remaining (a factor of 3). The time risk is also 3.

Out of a possible total of 12, the aircraft, environment, and pilot risks come to 9, which then has to be multiplied by the time risk of 3, to give a total potential risk of 27 out of a possible 36. One end of the beam is now pushed down by the weight of the 75 percent (27/36) risk factor. How can the safety beam be balanced? The pilot could (1) fly VFR, (2) choose an alternate within 30-minutes flying to remove fuel criticality, (3) reduce passenger tensions, (4) fly to an alternate destination in good weather and overnight.

Following a weather check that confirms that Houston is experiencing severe thunderstorms, the pilot decides to turn around. Weather at the alternate and fuel requirements are checked to complete the decision cycle by ensuring that the revised plan will work. By accurately assessing the risks, the pilot successfully concludes the flight. Arriving—not where you arrive—is the major criterion for assessing success in pilot judgment.

Discussion Questions (Part C)

1. Representatives of the Air Line Pilots Association, addressing the Second Symposium on Aviation Psychology at Ohio State University in 1983, pointed out what they felt were misconceptions about pilot judgment.

> Some people would have you believe that economics plays the key role in a pilot's decision. Others feel it is his psychological background and forces operat-

ing on his life. Still others believe the critical keys are "get-home-itis" or the subtle pressure of peers. We have little patience with people who believe these preconditions control the perception of information and cloud the mind of the pilot during decision making.

What, in your view, contributes to faulty pilot decisions? What can be done to rectify this situation?

Part D. Cockpit Resource Management

CREW PERFORMANCE OBJECTIVES

In QANTAS flight operations training, crew performance objectives are used when a number of people are required to achieve a specific task as a coordinated effort. For example, in the engine starting sequence on the B747, the captain, first officer, and flight engineer are all involved. Simultaneously, there has to be coordination with the tractor pushing the aircraft away from the terminal, monitoring of the aircraft movement, engine operations, systems that are activated, radio communications, and so on. Individual tasks exist in parallel with group tasks, thus training is required for both activities.

A focus has been made on cockpit resource management (CRM) as a means of efficient coordination of the material and human resources on the flight deck. Dr. John Lauber, NASA Ames Research Center, describes *cockpit resource management* as the effective use of available human, technical, and information sources.

One way of maintaining consistency in crew performance is to establish crew performance objectives. Objectives are expressed in behavioral terms, thus removing any ambiguity or conflict in expectations. For example, on finals an airspeed reference may be −0 or +5; distance off the glide slope of one dot, or plus or minus 16 feet at the other marker.

CRM involves a number of traditional means of crew coordination. Both industry and government have been working to facilitate an analysis of CRM through groups of experts such as the Cockpit Management and Crew Coordination Committee of the Flight Safety Foundation. For example, when a group of 13 aviation professionals met under the leadership of Dr. Richard Jensen of Ohio State University in March 1985, they distinguished between command (exercising authority through direction, implied intent, and example); management (coordinating the use of resources through planning, organizing, staffing, directing, and controlling); leadership (creating and maintaining an environment in which the crew can function effectively), and cockpit management, which was seen as the continuous process of integrating the human and technical information resources in the cockpit and outside the cockpit, with the primary goal of maximizing safety, efficiency and passenger comfort on each flight.

An emphasis was placed on two items: freeing lines of communication so that facts and opinions could be freely presented to the captain, and the importance of the initial briefing to establish openness to feedback from crew. To facilitate CRM training, the committee was able to identify twenty vital skills: listening, communication, assertiveness, awareness of the situation, ability to deal with conflict, problem solving, problem definition, priority and analysis, open-mindedness, personality awareness, managing distractions, fatigue management, use of checklists, decision making, pattern recognition, crew incapacitation recognition, workload assessment, division of attention, stress management, and ability to critique.

Methods of training discussed included line oriented flight training (LOFT) (using high-fidelity simulation of aircraft and air traffic control operations) and the use of video to enhance self-appraisal and the ability to critique. Role-playing and interactive learning through workbooks and computer assisted instruction (CAI) were also identified as useful instructional methods.

PRINCIPLES OF FLIGHT DECK RESOURCE MANAGEMENT

Another means of gaining efficient coordination of flight crew is the utilization of principles of flight deck resource management, such as those developed by American Airlines. They include

1. appropriate delegation of tasks and assignment of responsibilities,
2. establishment of a logical order of priorities,
3. continuous monitoring and cross checking of essential instruments and systems,
4. careful assessment of problems and avoidance of preoccupation with minor ones,
5. utilization of all available data to conduct an operation,
6. clear communication among crew members of all plans and intentions, and
7. assurance of sound leadership by the pilot in command.

The demands of critical flight events, such as an emergency, exceed the ability of individuals to cope. In multipilot operations, there has to be a consistent and accepted pattern of cooperation and contribution.

MANAGERIAL GRID®

Recognizing the body of research evidence supporting the claim that a group can collectively achieve more than the sum of individual performances (and more than any individual), the United Airlines approach to cockpit resource management has been one of teamwork with pilot acting as manager rather than director. Critical to the approach is the Blake and Mouton concept of leadership shown in Fig. 7.3. In this cockpit resource management grid, the vertical axis represents the concern for the individuals comprising the crew, and the horizontal axis represents the concern for task accomplishment. Given some 81 different locations for description of leader style, it is more fruitful to concentrate on the five coarse descriptors indicated. These represent the basic combinations of the two leadership dimensions initially identified in studies at Ohio State University in the forties and fifties. Additionally, they indicate that an effective leader is able to integrate skills along both dimensions so that square 9,9 is the highest position representing integrated leadership.

United Airlines and Scientific Methods, a company in Austin, Texas, that is involved in the design of airline leadership programs, have entered into a joint venture to provide cockpit resource management training to the aviation industry. The training materials that are used are based on the managerial grid and the learning technologies developed by Robert S. Blake and Jane S. Mouton. Combining these educational and theory-based approaches with United's simulator and technical training skills has led to an effective training program.

First, it is recognized that there is no panacea to the cockpit management problem and that achieving results is not simply a matter of gathering pilots into a room and providing a lecture on the importance of open communication or leadership behavior. Effective changes in teamwork skills cannot be realized simply by reviewing case studies of accidents and improving communication. Two central issues in determining how to best implement effective resource management training are (1) the content of what was to be learned and (2) the learning methodology to gain the needed behavioral changes. Results indicate that a sound, theory-based approach to the cockpit management problem can pay real dividends in improved safety through better teamwork performance.

THE GRID THEORY

The grid theory is a basis for providing the understanding of teamwork dynamics. Five elements have been identified as being important in a comparative system of learning cockpit management. These are inquiry, advocacy, conflict resolution, critique, and decision making. They provide a framework and a set of criteria, enabling anticipation and vigilance to replace complacency and assumption on the flight deck.

The Cockpit Grid uses a specially prepared text written for crewmembers. This is read as part of a home study program and is "Phase 1" of the

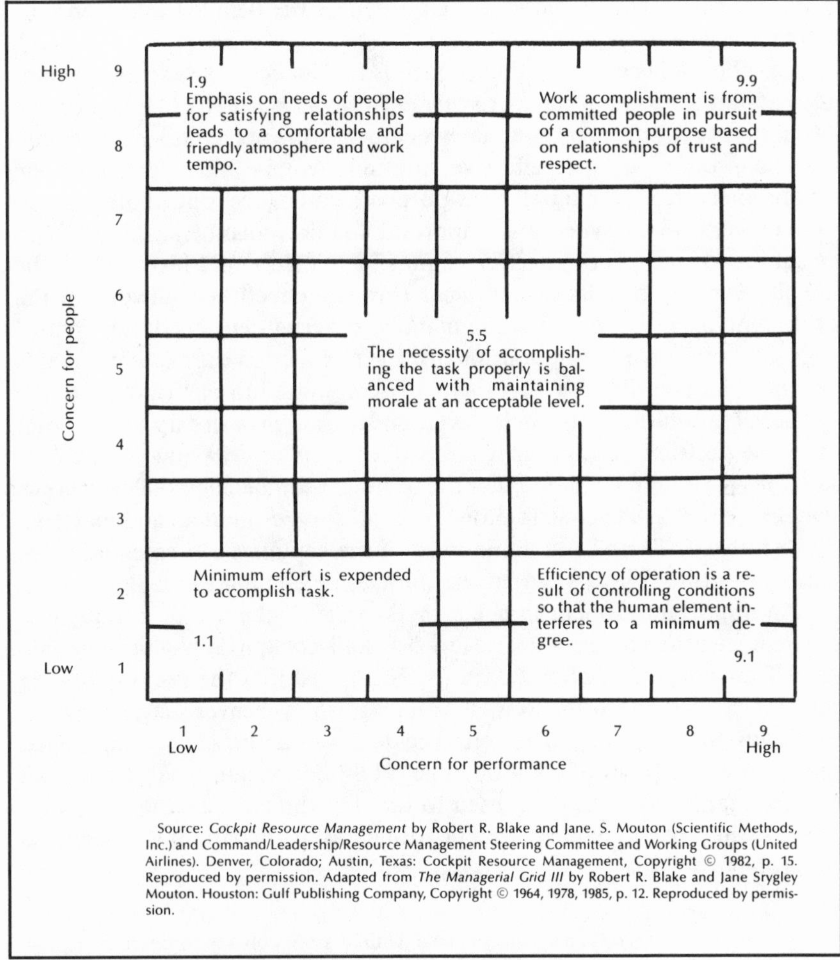

Source: *Cockpit Resource Management* by Robert R. Blake and Jane. S. Mouton (Scientific Methods, Inc.) and Command/Leadership/Resource Management Steering Committee and Working Groups (United Airlines). Denver, Colorado; Austin, Texas: Cockpit Resource Management, Copyright © 1982, p. 15. Reproduced by permission. Adapted from *The Managerial Grid III* by Robert R. Blake and Jane Srygley Mouton. Houston: Gulf Publishing Company, Copyright © 1964, 1978, 1985, p. 12. Reproduced by permission.

FIG. 7.3. **The Cockpit Resource Management Grid**

CRM training. It provides a common language for use in applying the succeeding phases of the program.

The second phase involves a structured learning process that allows crewmembers to learn, firsthand, how to use their newly acquired knowledge for understanding of behavioral effectiveness. Crewmembers are able to analyze how they react to various leadership styles in the cockpit and how their own behavior can affect operational outcomes. This phase of the training is conducted in a seminar environment, allowing the crewmembers to address this new area without the burden of the detailed attention normally necessary in flight.

Another focus of the grid approach is the concept of *synergy,* or combined action. The importance of synergy for aviation safety is based on the notion that two, three, or more crewmembers working together in a sound way can produce a more effective solution to a problem than any one person; and three working at cross-purposes can cancel one another out.

The concept of synergy is important for crewmembers, especially in the context of the use of effective command structure and hierarchy in the cockpit. Synergy and its achievement through effective teamwork in the cockpit enhances the captain's command and strengthens his or her control of the cockpit instead of diminishing it. Even in a crisis the captain remains open to input and the crew itself is jointly committed to contributing to the best possible solution. This holds even under time pressure up to the point where the decision becomes inevitable. This basis of crew interaction provides the maximum likelihood that the technical competencies of each crewmember, as well as the contributions from all sources, human and material, will be utilized. There is no diminution of the captain's ultimate authority, as it is not utilized in a way that discourages or eliminates input.

An important element in achieving synergy is the use of critique and feedback in order to optimize teamwork and cockpit behavior. The National Transportation Safety Board (NTSB) noted that the use of planning and effective critique prior to a B727 departure in Denver may have been significant in preventing a serious accident when the crew encountered windshear at the point of rotation. The NTSB commented that the cockpit resource management training used to develop the critique and teamwork skills for this crew may have been instrumental in contributing to its effective problem solving.

> The Safety Board believes that United's cockpit resource management training may have played a positive role in preventing a more serious accident from occurring in Denver and that it is an endeavor that should be encouraged. The Board previously has recognized the benefits of this training when it recommended in 1979, as the result of several accident investigations, in which the breakdown in cockpit resource management was identified as a contributing factor, that the FAA: "Urge . . . operators to ensure that their flightcrews are indoctrinated in principles of flight deck resource management, with particular emphasis on the merits of participative management for captains and assertiveness training for other cockpit crewmembers. (A-79-47)"

The self-study program and seminar make an indispensable contribution to better teamwork, but they can only be a part of the training if there is to be the hoped-for application in the flight environment. They do, however, provide a very strong foundation upon which to build for future operational effectiveness.

Having provided the opportunity for attaining an intellectual understanding of team dynamics, plus the opportunity to apply this understanding in a seminar environment, United has made cockpit resource management a part of its *recurrent* training program for all cockpit crewmembers. Thus, it is not presumed that a single input is sufficient to affect a lasting shift in behavior, attitudes, and values of crewmembers but that continuous and programmed follow-up is essential for long-term benefits to accrue.

Due to the success of the cockpit resource management training program, United has applied the principles to other areas of the flight crewmembers' training and supervision. CRM leadership concepts have been incorporated into the training that crewmembers receive when they upgrade from one cockpit crew position to another. In addition, these concepts have been made an integral part of both the enroute checks that are given annually and the initial operating experience subsequent to an upgrading exercise.

In addition, given the attention that is being directed to the need for improved cockpit resource management, several large air carriers have also instituted a resource management training program for their own flight crews. There have also been jointly sponsored seminars, in different parts of the world, to introduce this training; and branches of the armed forces in both the United States and Canada are engaged in the planning process to build CRM training into their training requirements. When this is added to the large number of corporations that have committed to send all their pilots through CRM training, it is clear that the overall aviation community recognizes the utility and need for this contribution to aviation safety.

SITUATIONAL LEADERSHIP

An alternate approach, which extends the idea of adapting leadership style to the situation, has been provided by Paul Hersey and Kenneth Blanchard of the Center for Leadership Studies in Escondido, California. Their situational leadership model has been used as the basis for the diagram shown in Fig. 7.4. This figure enables pilots in command to assess existing behavior in comparison with desired behavior. This can be done by estimating the maturity level of other crew, then projecting a line onto the "telling," "selling," "participating," or "delegating" cells. As well as the two dimensions of task directive behavior and crew relationship behavior, the leader must take into account (for the specific task to be performed) the maturity of the crew involved. For Hersey and Blanchard, *maturity* is the ability to set high but realistic goals, acceptance of responsibility, and the background of the individual or crew.

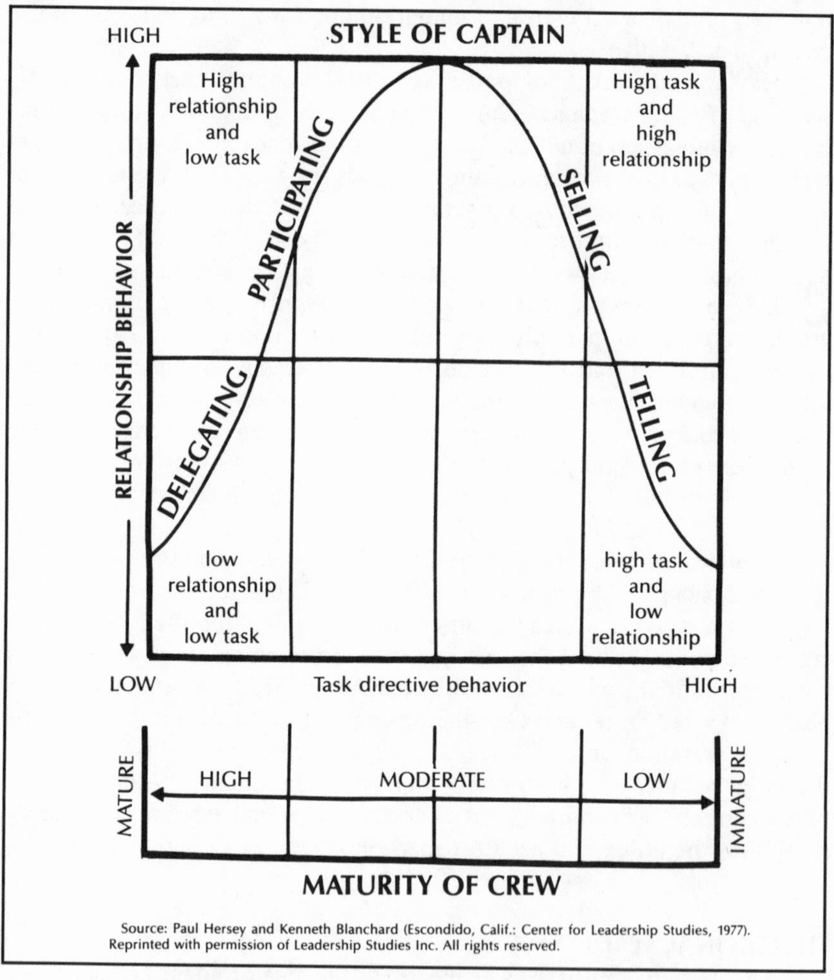

Source: Paul Hersey and Kenneth Blanchard (Escondido, Calif.: Center for Leadership Studies, 1977).

FIG. 7.4. **An Adaptation of the Hersey and Blanchard Approach to Situational Leadership**

This model has application to cockpit resource management and could provide a valuable core for the production of a seminar or workshop program to develop leadership skills in flight crew.

To a large extent, company standardization of procedures narrows the range of autonomy (and, thus, the responsibility for individual decisions) by flight crew. Such procedures have the additional advantages of assuring coordination, a common focus, and the elimination of superfluous talk. An example cited by a company check pilot was the "stop" frame of mind to

V1, and the "go" frame of mind when the pilot in command transfers concentration, after V1, from power to controls.

The same pilot, however, had a cautionary comment about the delegation of authority. Conscious that it is far easier to delegate authority than responsibility, he said his rule of thumb was, "Trust your crew to carry out their tasks, but do not rely on them." This caution has a reciprocal evidenced in cases in which copilot assertiveness was insufficient to prevent an accident due to captain incapacitation. The psychology of copilot takeover has been the focus of papers presented at recent human factors and aviation psychology conferences in the United States. Solutions suggested have ranged from a more obvious system of monitoring the performance of the pilot in command, to interactive training to counter the "macho" dominance syndrome, to self-awareness training.

Discussion Questions (Part D)

1. What is your definition of "cockpit resource management"?
2. In your view, how can that process be made more effective and more efficient?
3. On the Blake and Mouton Cockpit Resource Management Grid, as applied to leadership on the flight deck, how is the 9,9 position preferable to the 9,1 position?
4. In terms of airmanship and flight safety, isn't the 9,1 position preferable? Justify your answer.
5. Refer to the diagram (Fig. 7.4) indicating the application of the Hersey and Blanchard approach to situational leadership in flight. Provide one example of leadership behavior for each of the four styles: high task and low relationship (telling); high task and high relationship (selling); high relationship and low task (participating); low relationship and low task (delegating).

Further Reading

Bloom, B. S., et al., eds. *Taxonomy of Educational Objectives: Cognitive Domain.* New York: McKay, 1956.

Burns, R. S. *New Approaches to Behavioral Objectives.* 2d ed. Dubuque, Ia.: W. C. Brown, 1977.

Cruickshank, Donald R., et al. *Reflective Teaching Kit.* Bloomington, Ind.: Phi Delta Kappa, 1982. This kit is published in a loose-leaf folder to facilitate the removal and photocopying of lessons. There is no copyright on the contents to encourage such use. It would be an ideal source for in-house teaching workshops or instructor refresher courses.

Davidson, J. E. J. "Cockpit Resource Management Training." *Australian Air Pilot,* spring 1983, 19–24.

Diehl, A. Test Plan and Schedule for Judgement Training Demonstration Project in FAA Eastern Region. Washington: FAA, June 1985.

Flight Safety Foundation. *Summary of an International Industry/Government Workshop on Pilot Decision Making.* Arlington, Va. Nov. 1985.

Gagne, R. M., and L. J. Briggs. *Principles of Instructional Design.* 2d ed. New York: Holt, Rinehart and Winston, 1979.

Good, T. L., B. J. Biddle, and J. E. Brophy. *Teachers Make a Difference.* New York: Holt, Rinehart and Winston, 1975.

Jensen, R. "Pilot Judgement: Training and Evaluation." In *Pilot Error* edited by R. Hurst and L. Hurst. London: Granada, 1982.

Jensen, R. S., and J. Adrion. *Aeronautical Decision-Making for Instrument Pilots.* Columbus, Ohio: Aviation Research Associates, 1984.

Jensen, R. S., and R. A. Benel. *Judgement Evaluation and Instruction in Civil Pilot Training.* Washington: FAA/RD/78/24. 1977.

Krathwohl, David R., et al. *Taxonomy of Educational Objectives: Affective Domain.* New York: McKay, 1964.

Lawton, R., R. Clarke, and L. Benner. "Teaching Your Students to Balance Risks While Flying." *AOPA Flight Instructor's Safety Report* 12, no. 1 (Jan. 1986): 1–3.

Lester, L. F., A. Diehl, and G. Buch. *Private Pilot Judgement Training in Flight School Settings: A Demonstration Project.* Paper presented to the Third Symposium on Aviation Psychology, Ohio State University, Columbus, Ohio, April 1985.

Mager, R. *Preparing Instructional Objectives.* Palo Alto: Fearon, 1962.

Mané, A. "Airmanship: An Introduction." *First Symposium on Aviation Psychology.* Aviation Psychology Laboratory, Ohio State University. 1981. Technical Report APL-1-81, 161–65.

Proceedings of the Third Symposium on Aviation Psychology, edited by R. S. Jensen. Aviation Psychology Laboratory, Ohio State University, 1985.

Romiszowski, A. J., ed. *The System Approach to Education and Training.* London: Koga Page, 1970.

Roscoe, S. N. *Man as a Precision Resource: The Enhancement of Human Effectiveness in Air Transport Operations.* Savoy, Ill.: University of Illinois, Aviation Research Laboratory, 1974.

Telfer, R., and A. Ashman. *Pilot Judgement Training—An Australian Validation.* New South Wales: University of Newcastle, 1986.

Turney, C., et al. *Sydney Micro Skills.* Redeveloped versions. Sydney: University of Sydney Press, 1983.

8 Conclusion

The Importance of the Flight Instructor

A flight instructor is the greatest single factor affecting a student pilot's learning. There are ways that flight instructors can optimize their positive effect on students (and on aviation), as this book has shown. For example, a knowledge of learning and memory processes can help students to pay attention, and minimize any communication gap. Students can be helped to remember by instructors who present material in meaningful ways, and who show students how experienced pilots remember the procedure or information. Students can be helped to cope with anxiety and can be stimulated by a variety of teaching methods, media, and instructor styles to cope with the challenges of the flight syllabus. Appropriately planned, presented, and evaluated flying lessons are the characteristic of the effective flight instructor.

Educational research shows the following ingredients in teaching success:

- enthusiasm from instructors who like what they are doing and show it
- task-oriented or business-like behavior by instructors who know their function and purpose
- indirectness by instructors who take the time to draw information from students rather than provide it, or to question rather than lecture
- helping students by structuring questions and indicating the nature of the response that is expected or the means the student can use to obtain the answer
- using praise judiciously to reinforce student effort
- varying levels of discussion and questioning from simple factual matters to those requiring a substantiated or well-argued point of view

A convivial instructor providing feedback to a student working through a clearly defined sequence of tasks is a sound foundation. Add to that an ability to handle transitions in aspects of the program (or when the learning slows), a clear focus on the task, and an accommodation of individual differences in students, and the professional flight instructor is beginning to emerge.

Prof. Don Cruickshank at Ohio State University, after reviewing two eras of research, saw an effective instructor as "well-organized, efficient, task-oriented, knowledgeable, verbally fluent, aware of student developmental levels, clear (not vague), enthusiastic, self-confident, confident of student abilities, holding high expectations and a can-do attitude, friendly and warm, encouraging and supportive, attentive, accepting and tolerant."

Try that for a checklist!

Professionalization

Experience and insight have produced some expert and efficient flight instructors. Their accomplishments through the years have contributed to a reservoir of instructional alternatives that enable them to predict, with remarkable accuracy, the consequences of a particular approach to a certain student for a specific exercise. They have built their own body of learning and teaching theory. They have a professional base for their day-to-day activity.

For those instructors, this book should have provided a source of confirmation. It would have presented a new vocabulary as part of an explicit rationale for what was previously intuitive. For flight instruction to be professionalized, the appropriate theory has to be documented in a way that will enable the newcomer to acquire the technical basis through study, rather than years of experience. Both are necessary, of course, but neither the instructor, aviation, nor the student pilots awaiting can afford to allow experience alone to be the determinant of instructor proficiency.

This book started with a disclaimer. There could be no recipe approach to flight instruction, the reader was told. The book now concludes with a parallel caveat. Because we cannot definitively state *the* way to instruct in any given circumstance, it follows that we cannot advocate a single set of criteria from which to judge a flight instructor's performance. A single definition of good flight instruction is unavailable because we have found little agreement on what constitutes successful flight instruction. Some of the dynamics of effective teaching have been named, and these provide a means of assessing aspects of instruction. Student achievement is probably the best single measure we have. If, however, the task of assessing flight instruction is to be undertaken, it could best be done by using a multidimensional approach that considers student achievement and the means used to facilitate it.

There are a wide variety of successful styles of flight instruction. If this book has been successful, it will enable an instructor to be more consistently successful.

Further Reading

Cruickshank, Don. "Teacher Clarity." *Journal for Teacher Education,* Sept. 1985.

Good, T. L., B. J. Biddle, and J. E. Brophy. *Teachers Make a Difference.* New York: Holt, Rinehart and Winston. 1975. Chapter 4 provides research relating teacher behavior to student outcomes.

Hurst, R., and L. Hurst, eds. *Pilot Error—The Human Factors.* 2d ed. New York: Jason Aronson, 1982. The comprehensive coverage includes human factors, the role of education and training in accident prevention, PJT, flight-deck automation, terrain accidents, air traffic control, and research perspectives.

Perrott, E. *Effective Teaching: A Practical Guide to Improving Your Teaching.* London: Longman. 1982. Effective teaching, planning, presentation skills, questions, and analysis of classroom activities are discussed.

Roe, E., and R. McDonald. *Informed Professional Judgement—A Guide to Evaluation in Post-Secondary Education.* St. Lucia: University of Queensland Press, 1983. See especially pages 156–60 in which four categories of teaching (presentation, content, course management, and extra-curricular activities) are analyzed. Guiding principles to govern the evaluation of teaching are provided.

Glossary

Achievement motivation. Motivation for academic learning that is based on the ego-enhancement achieved by winning in a competitive situation: desire to achieve success, as opposed to desire to avoid failure.

Activation. See Arousal.

Advance organizer. A preinstructional strategy that involves prefacing information to be learned with a brief statement that organizes the concepts involved at a high level of abstraction.

Affective domain. The domain that emphasizes feelings, motivation, and reactions to other people and oneself; to be distinguished from the cognitive domain, which is concerned with thought.

Anxiety. A feeling of threat accompanied by high arousal. Trait anxiety is a predisposition for some individuals to react to a variety of situations with anxiety. State anxiety refers to the anxiety felt in a particular situation, such as test anxiety.

Arousal. Arousal, or activation, is a generalized motivation to behave, irrespective of the direction of behavior. Low arousal is associated with low drive; high arousal with "hyper" behavior. The physiological source of arousal is the reticular arousal system (RAS), which is located in the brain stem.

Assessment. In general, the evaluation of student performance, after a teaching episode. In progressive assessment, the student's final grade is determined by performance throughout the course; in terminal assessment, the grade is determined by performance in a final examination at the end of the course.

Associative interference. A theory of forgetting based on the similarity between items learned. Interference is retroactive when recall of previously learned material is inhibited by later material and proactive when recently learned material is inhibited by material that has been previously learned. This theory is mostly relevant for rote rather than meaningful learning.

Attending. Selecting relevant from irrelevant material in the sensory register through a process of precoding; being aware of material in working memory.

Attribution theory. A theory that bases motivation for performance of a particular task upon previous performance of that task. Specifically, attribution theory emphasizes the factors (such as ability, luck, effort, or task difficulty) to which the subject attributes his or her performance.

Behavioral objective. An educational goal that specifies the learned behaviors a student is to exhibit after a learning episode (lesson or series of lessons). The objective usually details the conditions under which the learning is to occur and the level of performance expected [see Objective (instructional)].

Behaviorism. A school of psychology that is concerned only with the conditions associated with changes in behavior. Internal (mentalistic or cognitive) events are regarded as irrelevant. Behaviorist psychologists rely on classical and operant conditioning as explanatory models of human behavior.

Chunks. The units of information grouped in working memory that determine the amount of information a person can handle at any given time.

Classical conditioning. See Conditioning.

Codes. Codes are the part of cognitive structure that allow input to be incorporated with previous learnings. Generic codes have a high degree of access to cognitive structure; surface codes access to only a few, limited aspects of cognitive structure.

Coding. See Encoding.

Cognitive domain. That aspect of human functioning that refers to thought; as opposed to the affective domain, which refers to feeling, emotion, and motivation.

Cognitive structure. The internal organization of codes or schemata that determines how information will be encoded. Cognitive structure generally grows more complex with development.

Cognitive style. A qualitatively distinct and consistent way of encoding, storing, and performing that is mostly independent of intelligence.

Conditioning. A form of learning (and of motivation) that suggests that the performance of responses is conditional upon the intervention of the external environment. In classical conditioning, responses are emitted as a function of the association between an unconditioned and a conditioned stimulus; operant (or instrumental) conditioning occurs when the responses are associated with rewarding or punishing consequences. Both forms of conditioning are examples of behaviorism.

Content learning. Learning in which the main interest of both teacher and student is the subject matter to be learned, rather than any effects the learning experience may have on the learner.

Contingency. A result (rewarding or punishing) promised prior to the elicitation of a response, in order to affect the frequency of that response.

Correlation. A statistical technique that expresses the degree of relationship between two variables: +1.00 indicates a perfect positive relationship (rarely obtained); −1.00 a perfect negative relationship (also rare); and 0.00 no relationship at all.

Criterion-referenced evaluation. Evaluation of student performance in terms of how well the student meets preset standards; as opposed to norm referenced evaluation.

Decoding. This term has two basic meanings: (1) retrieving material from long-term memory (see Dismembering), (2) translating the printed word into meaning as in reading text.

Dismembering. Based on the theory of memory, according to which individuals code the different aspects of an experience in semantic, temporal, logical, or spatial codes, and reassemble these components on recall.

Encoding. Reading in, interpreting, and understanding input (assimilation) in terms of existing coded knowledge; coding.

Entering behavior(s). A term for those behaviors (such as previous knowledge, skills, attitudes) that a student has developed before he or she enters a learning situation in which they are relevant to its successful completion.

Essay tests. Tests or examinations in which students structure their own response in continuous prose.

Evaluation. Sampling student performance and making a judgment as to its adequacy. Evaluation may be norm-referenced, criterion-referenced, formative, or summative.

External examination. A system of assessment in which exams are set and marked outside the teaching institution.

Extinction. The weakening and disappearance of a learned response through lack of reinforcement (e.g., in operant conditioning by withholding reward, and in classical conditioning by not periodically presenting the unconditioned stimulus with the conditioned stimulus).

Extrapunitive. An individual who blames others for the fact that he or she is punished for wrongdoing (as opposed to intropunitive).

Extrinsic motivation. Where learning or performance takes place as a means of gaining some material reward or avoiding a punishment, it is extrinsically motivated; learning undergone for material consequences.

Extrovert. An individual who generally operates with a lower-than-optimal arousal level, and who consequently feels more comfortable in surroundings likely to maintain high arousal (e.g., the company of other people).

Formative evaluation. Evaluation conducted during the performance of a task to provide feedback information on how well the task is being performed and how performance may be improved.

Individualized instruction. Instruction designed to meet the needs of the individual student in terms of content and method of teaching. In particular individualized, or personalized, systems of instruction (PSI) often mean just that the student proceeds at his or her own pace until the instructional objectives are met, as in mastery learning.

Information processing theories. A form of the cognitive theory of learning in which human behavior is described in terms of the individual selectively interpreting, storing, and recalling the information in the environment. This concept is analogous to that of a self-programming computer.

Intelligence. A hypothetical factor of wide generality that is presumed to underlie an individual's competence in performing cognitive tasks. A great deal of controversy surrounds the nature, generality, and modifiability of intelligence.

Intrinsic motivation. Where learning or performance takes place in the absence of any intrinsic, social, or achievement motivation, it is positively intrinsically motivated; where learning is abruptly terminated for no evident reasons it is negatively intrinsically motivated. Positive intrinsic motivation usually signals high-quality learning.

Intropunitive. Individuals who blame themselves for the fact that they are punished for wrongdoing (as opposed to extrapunitive).

Introvert. An individual who generally operates at a higher-than-optimal arousal level and who consequently feels most comfortable in quiet surroundings.

IQ. Intelligence quotient (IQ) is the score yielded by an intelligence test, which measures a person's general ability in relation to the population (see Intelligence).

Learning. The acquisition of skills or information through interaction with the environment.

Locus of control. A concept describing how individuals perceive themselves in relation to the external world. Individuals with an external locus of control perceive themselves as being controlled by chance, other people, or fate; those with an internal locus of control perceive themselves as having control over their decisions and what happens to them (see Pawn and Origin).

Long-term memory (LTM). Storage of previous learning to be reconstructed when appropriate in working memory.

Mastery learning. An individualized method of teaching and criterion-referenced evaluation, based on the assumption that virtually all students can learn basic core content if given sufficient time and adequate instruction.

Match-mismatch. The state in which input is to be interpreted by cognitive structure. Match implies complete interpretation (encoding or assimilation); mismatch the need for recoding or accommodation. According to the degree of mismatch, intrinsic motivation will be positive or negative.

Material consequences. The pleasant and/or unpleasant results of a particular behavior; these consequences are manipulated in behavior modification programs.

Mean. The average of a distribution of scores, that is, the sum of the scores divided by the number of scores.

Meaningful learning. Learning verbal material by the method of coding with the intention of understanding the message well enough to be able to express the sense of the message in different words.

Median. In a distribution of scores from lowest to highest, the median score is the one at the midpoint of the distribution. In a normal curve the median is the same as the mean.

Memory. The storage and retrieval of information. Three levels, based on period of retention, are postulated: ultrashort (sensory register), short (working memory), and long (long-term memory).

Mental set. A condition where the plan is consciously adjusted to attend to a particular task or train of information. Conscious attention to one line of activity over others.

Meta-cognition. Being aware of one's own mental processes themselves, rather than of the contents of those processes.

Mnemonic. An artificial device to aid memory by linking new information to well-known material. For example, A SEAL is a mnemonic for the prestall checklist (altitude, security check, engine check, area, lookout).

Modeling. Learning that takes place as a result of seeing someone else carry out the performance.

Multiple-choice. A form of objective test in which the "stem" of the item is presented, and the student has to select the correct response to the stem usually from four or five alternatives.

Need-achievers. Individuals who are highly achievement motivated.

Normal curve. A bell-shaped curve that is assumed to correspond to the distribution of many ability test scores.

Normalizing. Making a set of test scores conform to the properties of the normal curve so that scores on different tests may be directly compared.

Norm-referenced evaluation. Evaluation of student performance in terms of how well the student compares to some reference group such as class, age peers, and so on; as opposed to criterion-referenced evaluation.

Numeracy. Displayed competence in calculating correctly and in understanding and applying the four rules of number to real-life situations.

Objective (instructional). A precise statement of the teacher's intentions when designing a learning episode; of the various types of instructional objective, the behavioral objective is the most rigorous.

Objective test. A test format in which the student chooses from a limited number of alternatives to indicate a response. This obviates subjective judgment in scoring, though it does not necessarily eliminate bias in choosing the correct alternatives.

Orienting response. A response signifying that attention is being paid to a particular stimulus.

Origin. An individual with an internal locus of control who sees his or her behavior as being caused by his or her own decisions. De Charms links this belief with intrinsic motivation.

Pawn. An individual with an external locus of control who sees his or her behavior as being directed from outside himself or herself.

Percentiles. Rank ordering converted to a percentage basis; 99th percentile indicates a score at the top 1 percent of the population. The 50th percentile is the same as the median. A decile occurs every 10 percentiles, and a quartile every 25 percentiles.

Personality. Generally, the total pattern of cognitive and affective variables that give a person a characteristic stance toward the world. More specifically, personality variables are often meant to be those drawn particularly from the affective domain, and include introversion, extroversion, locus of control, neuroticism, and so on.

Plan. The "executive" system that controls and integrates cognitive processes from selective attention, coding and/or rehearsal, and storage and retrieval.

Posttest. In an experiment, a test conducted after the experimental treatment (as opposed to pretest).

Practicum. An instructional situation that involves practicing a role or learning a skill through active participation, for example, practice teaching in teacher education.

Precoding. The process, which takes place in the sensory register, of selecting a particular stimulus for conscious attention on the basis of its current importance.

Preinstructional strategies. Teaching strategies that attempt to focus the learner's attention on particularly relevant items of content before it is actually taught, for example, a list of questions, an overview, an advance organizer.

Premack principle. A principle suggested by David Premack to help the selection of suitable reinforcers in behavior modification. It states that if behavior *A* is more frequent than behavior *B,* the frequency of behavior *B* can be increased by making behavior *A* contingent on behavior *B* (see Contingency).

Pretest. In an experiment, a test conducted prior to any experimental treatment (as opposed to posttest).

Programmed instruction. Self-administered instructional materials in which a topic is broken down into small sequential steps (frames) that require a response from the learner. Immediate feedback to the learner is provided.

Punishment. An unpleasant event that follows a particular behavior as a result of a deliberate decision by an authority figure. Reactions to punishment may be intropunitive or extrapunitive.

Rank order. Placing scores in order from lowest to highest.

Recoding. Change in codes or cognitive structure brought about as a result of mismatch (see Match-mismatch). With the optimal degree of mismatch, cognitive structures grow more complex and intrinsic motivation is experienced.

Reconstruction. A theory of memory that postulates that recall of previously learned material is not so much a matter of retrieving stored information as reconstructing the original event from stored "clues."

Rehearsal. Storing material by repeated practice rather than by linking to previously learned material (encoding). In the case of verbal material, rehearsal produces rote learning.

Reinforcement. The process whereby a (rewarding) consequence of a response results in the increased likelihood of that response occurring in future. A positive reinforcer refers to a pleasant consequence; a negative reinforcer to the avoidance of an unpleasant consequence. Reinforcements may be allocated according to various schedules, such as partial, interval, self-administered, and so on.

Reliability. A characteristic of a test that is internally consistent (the items measure the same construct) and yields scores that are stable over time.

Remediation. Reviewing prior learning where it has been incorrect or otherwise inadequate and taking steps to rectify the faults.

Retrieval. The usual term for recalling previously learned material, although it may be misleading.

Rote learning. Learning verbal material by the method of rehearsal with the intention of exactly reproducing the original, with or without understanding it.

Selective attention. The process of attending to particular stimuli and not to others (see Precoding).

Self-concept. The image or concept one has of oneself, particularly of one's abilities (physical, mental, and social), and the value (positive or negative) one places on these self-evaluations.

Self-efficacy. A person's expectation that he or she will perform a task at a particular level of effectiveness. Beliefs in self-efficacy have a strong effect on intrinsic motivation.

Self-fulfilling prophecy. Occurs when believing or saying something is true actually causes it to happen; for example, not giving a student enough work in the belief that the student is stupid and "proving" the belief by the student's lack of output.

Semantic. Refers to word meanings: semantic input, or semantic coding, thus refers to verbal messages and to storing information according to word meaning. Semantic processing is believed to occur mainly in the left hemisphere of the brain.

Sensory input. Information supplied by the senses: mainly vision, touch, hearing, taste, and smell.

Sensory register. The first stage in information processing: a brief (about one second or less) period during which information is held and scanned (precoding).

Social motivation. Where learning or performance takes place because of the influence of one or more other people, as in modeling, or conforming to a group.

Spatial. Pertains to the arrangement of objects in space (e.g., a map). Spatial coding would refer to the storage of information in visual terms (e.g., an image). Spatial processing is believed to occur mainly in the right hemisphere of the brain.

Standardized tests. Tests that have been previously trialed and revised, which yield reliable (usually norm-referenced) information. The conditions of testing and scoring must be carefully adhered to.

Strategy. A way of tackling a type of problem or learning material, which may be applied to a whole class of learnings, not just to the particular problem in question. Individuals tend to select learning strategies according (in part) to their motives and desired outcomes.

Stress. Psychological or physical pressure that results in increases in arousal. Mild stress may improve performance, but heavy stress usually impairs performance (see Yerkes-Dodson Law).

Structure (of instructional settings). There are three aspects of structure: *instructional,* which minimizes the learner's decisions about learning; *motivational,* which involves extrinsic (positive and negative) reinforcers, social motivation, and achievement motivation; and *situational,* the provision of space and facilities for instruction.

Summative evaluation. Evaluation conducted after a task or learning episode has been completed in order to see how well it was done; grading a task. Summative evaluation may be norm-referenced or criterion-referenced.

Temporal. Refers to the relationships between events in time. Temporal coding is using chronological order as a basis for storing and remembering information, such as historical events.

Test. A task presented to the learner to assess the learner's performance; an important component in assessment and evaluation.

Test anxiety. A form of state anxiety that is focused on the test situation, and specifically on the individual's fear of failure. Test anxiety usually impairs performance.

Test distortion. Occurs when a particular test format, essay or objective, consistently inflates or depresses a person's performance because of some personality characteristic. For example, multiple-choice tests favor convergent thinkers over divergent thinkers when their content knowledge is identical.

Validity. The extent to which a test measures what it is designed to measure.

Variability (of teaching). Switching from one teaching medium, style, and content to another in order to maintain student interest.

Variability (of test scores). The extent to which a set of test scores varies from the mean of the group; usually expressed as standard deviation (SD) or variance (SD^2).

Working memory. The short-term memory system in which conscious thought takes place; roughly equivalent to "span of attention."

Yerkes-Dodson Law. A law of motivation formulated early this century stating that under increasing motivation (arousal), performance in complex tasks will be impaired before that in simple tasks. In the latter, performance may show improvement before impairment.

Index

Ab initio, 35, 42
Abstract, 9
Acceleration error, 32. *See also* Compass
Achievement motivation, 85–86, 151. *See also* Motivation
Advance organizer, 22, 151
Affective domain, 58, 151
Aim, 4, 5, 13
Airmanship, 134–35
Altimeter error, 32
American Airlines, 139
Ansett Airlines, ix
Anxiety, 67, 82, 108, 147, 151
  state, 67
  test, 69
  trait, 67
AOPA, 136, 137
Approach to landing, 35, 36, 45–48
  crosswind, 29, 35
Arousal, 59–68, 75–76, 151
  energizing and interfering effects, 60–61
  information from senses and cortical processes, 61
  orienting response, 60
  and performance, 62–65
  and task complexity, 65
  and variability, 71
Assessment, 151. *See also* Evaluation
Associative interference, 40, 41, 151. *See also* Forgetting
Attending, 151. *See also* Attention
Attention, 14, 15, 18, 20, 39
Attitudes, 9
  pilot, 13
  pilot judgment training, 134–37
Attribution theory, 90–93, 152. *See also* Motivation
Autonomous level, 48

Behavior, 80–92. *See also* Motivation
  consequences, 81–82
Behavioral objective, 21, 120–28, 152. *See also* Objective
Behavioral verb, 127. *See also* Objective
Bernoulli's Theorem, 27–28

Center for Leadership Studies, 143
Center for Vocational Education, 13
Check and training, 125
Clarity, 133
Closed loop, 48
Cockpit resource management (CRM), 138–39
Code, 15, 16, 38, 39, 152
  coding, 16, 70, 71
    logical, 39
    semantic, 38
    spatial, 39
    temporal, 38
  dismembering, 39
  generic, 70
Coding, 152. *See also* Code
Cognitive
  domain, 58, 152
  level, 45
Communication, 139
Communication gap, 19–20
Compass, 30
  acceleration error, 32
  deviation, 30
  leading, 30, 31
  turning error, 32
  variation, 30
Computer assisted instruction (CAI), 139
Concentration, 18
Concrete, 9
Conditioning, 79, 152. *See also* Motivation
Consolidation, 71. *See also* Learning
Content learning, 152. *See also* Flight instruction, elements
Contract system, 115–16. *See also* Evaluation

Conversion, 26
Cortical processes, 60–61
Crew performance objectives, 138–39
Criterion-referenced evaluation, 98–100, 107, 108, 153. *See also* Evaluation
Crosswind. *See* Approach to landing

Decision making, 134–38. *See also* Pilot judgment training
Deviation. *See* Compass
Directing attention. *See* Attention
Dismembering, 39, 153. *See also* Code
Downwind, 33

East-West Airlines, ix
Efficacy, 90. *See also* Attribution theory
Elaboration, 70
Embry-Riddle Aeronautical University, 135
Emphasis, 133
Essay, 104–6, 155. *See also* Tests; Evaluation
Evaluation, 94–118, 153
- comparison of tests, 112
- constructing tests, 108–11
- criterion-referenced, 99
- essay, 104–6
- formative and summative, 95
- grading, 115
- measurement error, 102
- normal curve, 96–97
- normalizing scores, 96–97
- norm-referenced, 95
- objective test, 106
- oral test, 108
- practical, 107–8
- quantitative, 101
- reliability, 101
- short-answer test, 106–7
- standard deviation, 96–97
- tests, 94–95
- validity, 103–4

Examples, 133
Extrinsic feedback, 50, 55
Extrinsic motivation, 50, 79, 153. *See also* Motivation

FAA, 40, 135
Feedback, 49, 133
- extrinsic, 50, 79, 153
- intrinsic, 50, 55

Fixative level, 45
Flight deck resource management, 139
Flight instruction, ix, 147–49
- case study, 3–10
- elements, 10–12

Flight instructor course, ix
Flight instructors, vii, 147–49
Flight Safety Foundation, 138, 146
Forgetting, 40, 41, 54
Formative evaluation, 95. *See also* Evaluation
Front, 33

Generic code, 27, 28, 70. *See also* Code; Learning
Grading, 115–16. *See also* Tests; Evaluation

Heading. *See* Compass

Imprinting, 55. *See also* Skills
Input, 39
Inserted question, 22
Instructional objective, 109, 120–28. *See also* Objective
Instructional strategies, 29, 132–33. *See also* Teaching
Instructional technique, 5, 11
Instrument rating, 99
Intelligence, 36, 154
Interaction of instructor and student, 73–74. *See also* Variability
Interference, 75
Internal state. *See* Physiological state
Interview, 108. *See also* Tests; Evaluation
Intrinsic feedback, 50, 55
Intrinsic motivation, 86, 154. *See also* Motivation

Knowledge, 9

Latitude. *See* Navigation
Leadership, 140–45
Learning, 16, 21, 22, 24–43, 70–71, 154
- consideration, 71
- elaboration, 70
- generic, 27–28
- meaningful, 24–43
- mistakes, 28
- mnemonic, 5, 29, 41
- organization, 71
- rate, 24–43
- recoding, 25–26
- space-saving strategies, 29–30
- surface, 27–28

Line oriented flight training (LOFT), 139
Logical coding. *See* Code

Longitude. *See* Navigation
Long-term memory, 154. *See also* Memory

Managerial Grid, 139–43. *See also* Flight deck resource management
Marker beacon, 34
Massed versus distributive practice, 53, 54. *See also* Skills
Mastery learning, 116, 154. *See also* Learning
Meaningful learning, 155. *See also* Learning
Measurement error, 102. *See also* Evaluation
Media and materials of instruction, 73. *See also* Variability
Memory, 6, 16, 22–37, 37–43, 59, 155
  long-term, 37–43, 59
  ultrashort-term, 16, 17
  working, 16, 22–37, 53, 59
Mental set, 18, 19, 155
Mistakes. *See* Learning
Mnemonic, 5, 9, 29, 41, 155. *See also* Learning
Modeling, 51, 56, 84–85, 155
Motivation, 20, 58, 78–91
  achievement, 85–86
  attribution theory, 90–93
  conditioning, 79–81
  creating optimal conflict, 86–87
  extinction, 82
  extrinsic, 79–88
  intrinsic, 86
  negative reinforcement, 82
  postive reinforcement, 81–82
  punishment, 83
  reward, 87
  self-concept, 84–85
  self-efficacy, 88–89
  self-reinforcement, 84
  social, 84–85
  timing, 83–84
  verbal persuasion, 89–90
  wanting to learn, 87

NASA, 138
National Association of Flight Instructors, ix, 20, 45
National Transportation Safety Board, 142
Navigation
  heading, 30, 31
  latitude, 31
  longitude, 31
Needs, 85–86, 155
  need-achievers, 85
Negative reinforcement, 82. *See also* Motivation
Nonverbal variations, 72–73. *See also* Variability
Normal curve, 97, 155. *See also* Evaluation
Normalizing scores, 96–97, 155. *See also* Evaluation
Norm-referenced evaluation, 95–115. *See also* Evaluation

Objective, 13, 21, 45, 155
  behavioral, 21
  behavioral verbs, 127
  criticisms, 122–24
  for and against, 122
  instructional, 109
  using objectives in flight instruction, 121
Objective test, 106, 155. *See also* Tests; Evaluation
Occluded front. *See* Front
Ohio State University, 137, 138, 146, 148
Operations training, 74–75
Oral test, 108. *See also* Tests; Evaluation
Organization, 71. *See also* Learning
Origin, 89, 156. *See also* Motivation
Overload, 35
Overview, 21, 22

Pacing, 87
Pawn, 88, 156. *See also* Motivation
Perceiving. *See* Perception
Perception, 20, 21
Persuasion, 89. *See also* Motivation
Physiological state, 19
Pilot judgment training, 134–37
Plan, 59, 156
Positive reinforcement, 81–82. *See also* Motivation
Posttest, 21, 156
Practical test, 107–8. *See also* Tests; Evaluation
Precoding, 16, 19, 20, 59, 156
Preflight brief, 4
Preparation, 113. *See also* Evaluation
Pretest, 21, 156
Processing, 14, 15
Professionalization, 148–49
Professionals, ix, 148–49
Progressive-part method, 52. *See also* Skills

Qantas, ix, 74, 75
Quantitative evaluation, 101–2. *See also* Evaluation
Questioning, 62

Recipe approach, vii, viii, 11, 148
Recoding, 157. *See also* Learning
Reconstruction, 40, 157. *See also* Forgetting
Reflective teaching, 129–30. *See also* Teaching
Rehearsal, 15, 39, 41, 49, 157. *See also* Skills; Remembering
Reinforcement, 81, 157. *See also* Motivation
Reliability, 101, 157
Remembering, 39
  rehearsing, 39
Retention, 17
Reverse-part method, 52. *See also* Skills
Reward, 87. *See also* Motivation
Risks, 136–37. *See also* Pilot judgment training
Rote learning, 157. *See also* Learning

Sample item, 22
Scientific Methods, 140, 141
Selecting, 6, 157
Self-concept, 84–85, 157. *See also* Motivation
Self-discipline, 83. *See also* Self-reinforcement; Motivation
Self-efficacy, 88–89, 157. *See also* Motivation
Self-evaluation, 115. *See also* Evaluation
Self-reinforcement, 84. *See also* Motivation
Semantic coding, 158
Sensory register, 15–18, 59, 158
Short-answer test, 106–7. *See also* Tests; Evaluation
Simulator, 9, 54
Situational leadership, 143–44. *See also* Leadership
Skills, 44
  autonomous level, 48
  cognitive level, 45
  conditions for learning, 49–50
  decay, 54
  fixative level, 45
  flying, 9, 44–56
  massed versus distributive, 53, 54
  for teaching, 10–12, 125–27, 129–33, 147–48
  whole versus part, 51, 52
Social motivation, 84–85, 158. *See also* Motivation
Solo flight, 19, 82, 89
Space-saving strategies, 29, 30. *See also* Learning; Teaching
Spatial coding, 158. *See also* Code
Stages of learning, 14
Stalling, 4–10
Stalls. *See* Stalling
Standard deviation, 96–97. *See also* Evaluation
State anxiety. *See* Anxiety
Stimulus, 18
Storing, 15, 16
Stress, 135–36, 158
  origins and effect, 63
  pilot workload, 66
  student, 68
Student pilots, 78–93
Summative evaluation, 95, 158. *See also* Evaluation
Surface code, 27, 28. *See also* Learning
Syllabus, 12
Synergy, 142. *See also* Flight deck resource management

Taxonomies, 13, 128. *See also* Objective
Teacher education, 13
Teaching, viii, 13
  reflective teaching, 129–30
  skills, 132–37
  space-saving strategies, 29–30
Teaching skills, 132–33. *See also* Teaching
Temporal coding, 158. *See also* Code
Test anxiety, 69, 158. *See also* Anxiety
  coping with, 69
Testing, 104–16, 158. *See also* Tests, Evaluation
Tests, 94–95, 101–2, 104–16, 158. *See also* Evaluation
  comparison, 112
  constructing, 108–11
  coverage, 113
  essay, 104–6
  mastery learning, 116
  objective, 106
  oral, 108
  Pass/Fail, 116–17
  practical, 107–8
  preparation, 113
  short-answer, 106–7
Theory, 11
Timing, 83–84. *See also* Motivation
Trait anxiety. *See* Anxiety
Transport Canada, 135
Turning error, 32. *See also* Compass
Turn maneuvers, 124

Ultrashort-term. *See* Memory
United Airlines, 139, 140
University of Illinois, 134
Upwind, 33
U.S. Federal Aviation Administration, 135

Validity, 103–4, 159. *See also* Evaluation
Variability, 71–74, 159
  interaction, 73–74
  media and materials, 73
  nonverbal, 72–73
  verbal, 72
Variation. *See* Compass; Input
VASI, 34
Verbal variations, 72. *See also* Variability
"V" Speeds, 31

Whole versus part teaching, 51, 52. *See also* Skills
Williams Air Force Base, 53
Working memory, 16, 43, 53, 159. *See also* Memory
Workload, 66
Wrong Stuff, The, 42

Yerkes-Dobson Law, 64, 159

Author Index

Adams, J. A., 54, 57
Adrion, J., 146
Ashman, A., 135, 146

Benel, R. A., 146
Benner, L., 136, 146
Biddle, B. J., 146, 149
Biggs, J. B., 43, 76, 93
Blake, R. R., 141, 145
Blanchard, K., 143, 145
Bloom, B. S., 145
Briggs, L. J., 146
Brophy, J. E., 146, 149
Buch, G. de B., 146
Burns, R. S., 145

Childs, J. M., 118
Clarke, R., 136, 146
Cruickshank, D. R., 145, 148, 149
Curzon, L. B., 13, 56

Davidson, J. E. J., 145
De Charms, R., 88
Detroi, A., 45–48
Diehl, A., ix, 135, 145, 146

Gagne, R. M., 52, 118, 146
Gainer, C. A., 54, 57
Glasser, W., 83
Good, T. L., 146, 149

Heckman, R., 40
Hersey, P., 143
Hockey, R., 77
Holmes, T. H., 63
Hurst, R. and L., 149

Jensen, R., ix, 135, 138, 146

Krathwohl, D. R., 146

Lauber, J., 138
Lawton, R., 136, 146
Lester, L. F., 146

McDonald, R., 149
Mager, R., 146
Mané, 134, 146
Mengelkoch, R. F., 54, 57
Mouton, J. S., 141, 145

Nance, J. J., 57

Perrott, E., 149
Premack, D., 82

Rahr, R. H., 63
Roe, E., 149
Romiszowski, A. J., 146
Roscoe, S., vi, ix, 51, 57, 118, 146

Telfer, R. A., 43, 76, 93, 135, 146
Thorndike, R. L., 118
Turney, C., 77, 146

DATE DUE

DEC 1 6 '98			
APR 1 3 1999			